COMPLEXIFICATION

COMPLEX

IFICATION

Explaining a Paradoxical World Through the Science of Surprise

JOHN L. CASTI

HarperCollins*Publishers*

Copyright acknowledgments follow page 318.

HarperCollins books may be purchased for educational, business, or sales promotional use. For information please write: Special Markets Department, HarperCollins Publishers, Inc., 10 East 53rd Street, New York, NY 10022.

FIRST EDITION

Designed by Irving Perkins Associates

Library of Congress Cataloging–in–Publication Data

Casti, J. L.

Complexification : explaining a paradoxical world through the science of surprise / John L. Casti.—1st ed.

p. cm.

Includes index.

ISBN 0-06-016888-9

1. Chaotic behavior in systems. 2. Paradox. I. Title.

Q172.5.C45C38 1994 93-36049

003.7—dc20

94 95 96 97 98 CC/HC 10 9 8 7 6 5 4 3 2 1

CONTENTS

PREFACE

When asked if ants had common sense, Edward O. Wilson, the famed Harvard entomologist, two-time Pulitzer prizewinner and father of sociobiology, replied: "If common sense means living by a set of rules of thumb that have worked well in the past, but living without examining those rules too closely or in detail, then, yes, ants have common sense." Surprise is what happens when common sense fails. Wilson's criterion emphasizes the point that what's involved in a surprise is intimately tied up with the idea of acting in accordance with a set of rules. When the rules of reality generating the events of daily life part company with the rules of thumb built upon everyday common sense, surprise is the outcome. This is a book about those rules of reality, their form, nature and idiosyncrasies. In particular, it is a book examining the degree to which we can aspire to the creation of a "science of surprise."

In everyday parlance, the word *surprise* represents the difference between expectations and reality; the gap between our assumptions and expectations about worldly events and the way those events actually turn out. In essence, surprises are the end result of predictions that fail. So to address the root causes of surprise, we're inevitably forced to look at how we go about making predictions.

Normally, predictions are made by following some kind of rule. Here's an example: "I predict that the Dow Jones Industrial Average will be higher at the end of the year than at the beginning if a team from the old National Football League is the winner of the Super Bowl game in January." This is the so-called Super Bowl Indicator for stock price fluctuations on the New York Stock Exchange.

The Super Bowl Indicator is an example of an explicit rule for making a prediction. Such rules are exemplified by doing a calculation. But there are also implicit rules, where we follow a hunch or make an intuitive leap of faith. In these cases, the cognitive rule we're using may be unknown to us even as we employ it. But in either case, if the rule we employ proves useful enough over a long period of time, we often give it a name, calling it something like *a rule of thumb, conventional wisdom, an old wives' tale* or just plain *common sense.* Surprises occur when following such rules leads to predictions that fail to match up to reality. Our task in this volume is to look at why the rules obeyed by nature and man have a disconcerting tendency to differ from those we've learned to use over several lifetimes' worth of experience. Now let me discuss briefly the structure of the book.

Somewhere in the middle of the film *Casablanca,* Rick and Ilse have the following exchange:

> ILSE: *Can I tell you a story, Rick?*
> RICK: *Has it got a wow finish?*
> ILSE: *I don't know the finish yet.*
> RICK: *Well, go on, tell it. Maybe one'll come to you as you go along.*

I felt much the same way while writing this book. And, like Ilse, I still don't know the finish. But no matter. Like most journeys, the trip is far more interesting than the destination.

It's often been noted that the rules of reality are subtle—but not malicious. These subtleties, however, take many forms, each of which may give rise to its own characteristic brand of surprise. This volume addresses several surprise-generating mechanisms responsible for our inability to make good predictions about what the systems of nature and humankind are going to do next and why. Chaos is one such mechanism, but there are others—catastrophes (instability), uncomputability, irreducibility and emergence—each of which is the focus of a chapter of this book. Within each chapter we examine the system-theoretic reasons why we will never be able to eliminate surprise from our lives, yet can still hold out the possibility of creating something approximating a science of the surprising.

It's my hope the book will underscore the fact that systems displaying surprising (i.e., unpredictable) behavior are more or less synonymous with those we regard as being in some way "complex." So the reader should see the book as both an exposition of the science of surprise and an introduction to the mysteries and peculiarities of complex, as opposed to simple, systems.

In my earlier volumes, *Paradigms Lost* and *Searching for Certainty,* I learned the value of comments, complaints and a frank exchange of views from readers. So again I warmly encourage readers who feel the need to drop me a note with their reactions to the ideas presented here. These remarks should be addressed to me c/o Santa Fe Institute, 1660 Old Pecos Trail, Santa Fe, NM 87501, USA.

JC
Santa Fe, 1993

ACKNOWLEDGMENTS

As always, the generous advice of friends and colleagues has been crucial in helping me bring this book to press with a minimum of grammatical, technical and conceptual faux pas. Special accolades in this regard go to Atlee Jackson and George Johnson, both of whom generously squandered their time in reading each and every word of the manuscript, and who commented extensively on my slovenly literary and scientific ways. In short, they acted as "guardian angels" for the reading public. Without their sterling efforts, this would have been a rather different book—and not for the better.

For administrative and organizational services par excellence, it's always a pleasure to acknowledge my literary agent, John Ware, for his efforts in attending to the many details needed to bring a book to market. In this same regard, but on the editorial side of the house, Larry Ashmead and Eamon Dolan at HarperCollins have provided the kind of editorial guidance that protects an author from himself—especially one trying to follow that thin line between presenting enough detail to make the material informative and leaving out enough detail to make the book comprehensible to the general nonscientific reader.

Let me also take this opportunity to acknowledge helpful discussions with many other friends and colleagues about matters scientific and otherwise. In no particular order, these hardy souls include Gustav Feichtinger, William Brock, Blake LeBaron, Brian Arthur, Manfred Deistler, Harold Morowitz, Murray Gell-Mann, Stuart Kauffman, Steen Rasmussen, Ake Andersson, Anders Karlqvist, Robert Rosen, Chris Langton, Tom Ray, Jeff Johnson, Mike Simmons, Ed Knapp, Bruce Abell, Ed Angel, Bernard Moret, Greg Chaitin, Ed Thorp, Karl Sigmund, Peter Weibel, Paul Makin, Bruce Sawhill, Doyne Farmer, Norm Packard, George Cowan, Martin Shubik, Mitch Waldrop, Chris Barrett, Lucien Duckstein, Sid Yakowitz, Harald Atmanspacher, Werner Schimanovich, Mel Shakun, Roland Thord, Dave Batten, Reuben Hersh, Joe Traub, and Phil Davis.

COMPLEXIFICATION

ONE

THE SIMPLE AND THE COMPLEX

Realities, Rules and Surprises

Trying to define yourself is like trying to bite your own teeth.

—ALAN WATTS

The medium is the message.

—MARSHALL MCLUHAN

Some problems are just too complicated for rational, logical solutions. They admit of insights, not answers.

—JEROME WIESNER

IN THE BEGINNING IS THE WOR(L)D

A few years ago I made an extended lecture tour of Japan. Due to the press of other responsibilities, my wife was unable to accompany me on this trip. This sad fact happened to come up one night at a dinner party shortly before my departure, whereupon one of our friends remarked, "Well, dear, you know what they say. Absence makes the heart grow fonder." By way of counterpoint, one of the other guests immediately stepped in with the rejoinder, "But they also say, 'Out of sight, out of mind.'" So what kind of sense—common or otherwise—can be squeezed out of either of these mutually contradictory proverbs? Or are they completely useless as far as offering a glimmer of insight into how my wife or I were likely to feel during that period of prolonged separation? In 1986, British psychologist Karl Teigen concocted an ingenious experiment to try to find out.

Teigen's experiment involved taking twenty-four well-known proverbs and transforming each of them into its opposite. So, for example, the proverb from the dinner party, "Out of sight, out of mind," became "Absence makes the heart grow fonder." Teigen then gave his students lists containing some genuine proverbs intermingled with those he had just formulated by this mirror-imaging process. He then asked the students to rate the proverbs for both originality and truth value. If you've already guessed the results, then you're just about half a step ahead of me. As one might have expected, the students could find no recognizable difference between one set of eternal truths and their opposites. In short, almost any observation—or its opposite—can be taken as a pithy encapsulation of everyday, garden-variety common sense.

In a set of related observations, the editor and writer Edward Tenner recently asked, Why do the seats seem to get smaller as the airplanes get larger? And why does voice mail seem to double the time it takes to complete a telephone call? Tenner's answer is what he calls the "revenge effect," the process by which an indifferent nature seems to get even with us lowly humans by twisting our cleverness back against us. While facing unintended consequences from our mucking about as we try to make things better is hardly a new phenomenon, Tenner notes that technology has magnified the effect to the point where we now have to weigh the potential consequences of our actions more carefully than ever before.

Both of these stories illustrate ways in which common sense seems to lead us to conclusions that are at variance with the way we *think* the world should work—ways based on beliefs we create from a lifetime's worth of observations and experience. The problem is that there's often a Grand Canyon–sized chasm between what we think and what is actually the case. And it's when this gap becomes too large for comfort that science and its many competitors in the reality-generation game—religion, art, music, mysticism—enter the scene. In all cases, what we're looking for is some way to compress our observations into a small set of easily digestible rules, or models, that will serve as guides to what we should expect and what we should do in an increasingly complex, hard-to-understand world.

Our goal in this book is to look at various types of complicated structures and counterintuitive behaviors, keeping an eye out for common features that could form the basis for a "science" of the surprising and the complex. Of course, in everyday terms the word *surprise* is just shorthand for the way we feel upon discovering that our pictures of reality depart from reality itself. So the crux of the problem of understanding the origin and nature of surprise resides in asking how those pictures of reality are formed in the first place. In science, the chosen way to paint a picture of reality is to build a model, often expressed in the compact language of mathematics. We try to encode our experiences of the real world into the symbols and rules of a mathematical formalism, and then make use of this formalism to generate predictions of what will transpire in the future. So from a scientific point of view, surprise can arise only as a consequence of models that are unfaithful to nature.

But models can fail to be true to reality in a number of inequivalent ways. They can display complicated, chaotic behavior, making it difficult to use them to generate accurate predictions; they may be unstable, so that minuscule changes in one part of the model give rise to very large changes in the model's predictions; they may involve quantities that are just plain uncomputable, even in principle; they may involve hard-to-understand linkages among the various parts of the model, connections that prevent us from analyzing the behavior of the model by breaking it into smaller, more digestible pieces. It's surprise-generating mechanisms like these that constitute the multiple foci of our deliberations in this volume. Everyone will admit that surprises can and often do occur; this much is pretty obvious. But here we set our sights a bit higher, trying to understand *why* surprises occur and whether or not there's anything we can do about them. Moreover, it's a *scientific* understanding of the ways and whys of the world that we're after, not merely anecdotal accounts. So with these goals in mind, the natural starting point for our attack on surprise and "complexification" is at the level of observations themselves and how these observations get transformed into measurements and models.

Observations, Numbers and Models

As it's taught in courses on the philosophy of science, the scientific answer to a question is a set of rules, or as such rules are more commonly termed, a model. To illustrate, suppose you ask why a mixture of two parts hydrogen and one part oxygen forms the clear fluid we call water. A scientific answer invokes a variety of rules from chemistry and physics describing the molecular structure of hydrogen and oxygen, together with another set of principles (or rules) for how two atomic structures fit together to produce a compound substance. And if we are so bold as to introduce symbols like H and O to represent a hydrogen and an oxygen atom, respectively, the end result of these rules can be compactly written down in symbolic form as $H_2 + O \rightarrow H_2O$.

This expression is a particularly crude form of what might charitably be called a *model* to explain the formation of water from hydrogen and oxygen. But note that the starting point of the scientific explanation was the wholly unexpected and seemingly unlikely observation that by combining two rather active gases, we can form not only a liquid but a liquid whose properties differ radically from the properties of either of its constituent parts. A detailed model of this reaction then gives an explanation for this surprising, counterintuitive outcome. Moreover, the set of rules constituting this model might also provide us with a systematic procedure for predicting what would happen should we attempt to combine another pair of gases in a similar way. Note carefully in the above setup that the starting point of the whole exercise was the crucial observation that hydrogen and oxygen can be combined in a certain way to form water. Without this observation, there's no place for science or any other reality-generation mechanism to get a foothold.

So what do we mean by an observation? In everyday parlance, an observation is just the memory trace left behind in our brains when the outside world impresses itself upon us via our sensory channels of sight, sound, touch, smell or taste. So, for instance, the sickly sweet smell of burning incense, the tangy taste of barbecued spareribs and the blinding flash of a lightning bolt all qualify as observations. In science we usually try to code these memories by numbers, mostly for the sake of compactness and so that we will have a common scale by which to compare different observations. This kind of coding also has the salutary side effect of providing the basis for representing the observations in symbols, hence allowing us to encode the world in stylized mathematical terms.

But as anyone who's ever tried carrying out such quantification knows, the translation of an observation into a measurement is a delicate and subtle matter indeed. And, in fact, the transition from sights and sounds to symbols and numbers is one that those of a humanistic bent claim cannot be meaningfully carried out beyond the cozy confines of a physics lab or, perhaps, the floor of the stock exchange. I'll leave it to the reader to judge the validity of these claims as we proceed. For now, let it suffice to say that there's a lot of misunderstanding on both sides of the humanist/scientist divide regarding this point (and a lot of other points as well). I hope a clearer picture of both the possibilities and limitations of quantification will emerge over the next few hundred pages.

To see how a model helps us understand what would otherwise be completely mystifying behavior, consider a pyramid of apples of the sort on display at any streetcorner fruit-and-produce stand. Suppose we ask which of the apples in the stack is under the greatest pressure. Common sense would probably suggest that it's the apple at the center of the bottommost layer of the pyramid. Let's see.

In 1981, Czech scientists J. Schmid and J. Novosad used sensitive pressure gauges to measure the forces acting at the base of a pile of sand particles. They discovered the surprising fact that the pressure is not greatest on the central point of the base at all, but on a ring of particles some distance from the center of the base. So in our pyramid of apples, the apple at the center of the base is under the *lowest* pressure of all the apples in the bottom layer. The overall pressure-versus-height relationship discovered by the Czech researchers is shown in Figure 1.1.

Recently, Kurt Liffman at the University of Melbourne developed an "electronic sandpile" inside his computer that attempts to mimic the kind of real-life sandpile studied by Schmid and Novosad. Liffman's set of rules (that is, his program) for the sandpile duplicates most of the forces seen in the real pile. But as yet it doesn't explain

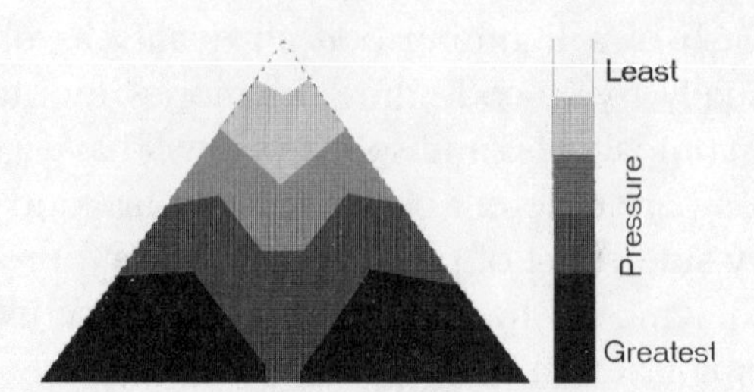

FIGURE 1.1. PRESSURE LEVELS IN A SANDPILE

the central dip in downward pressure. So the model is a halfway house toward understanding the properties of a sandpile. Clearly, there are still rules of sandpile behavior needed to supplement those forming Liffman's model. And whatever these unknown rules turn out to be, industry definitely wants to know about them since it's of considerable economic importance to be able to predict when a pile of iron ore or fertilizer may cake or fragment under its own weight. So here we have a good illustration of the power—and the limitations—of mathematical and computer models of natural processes. The computer model predicts and explains some of the important features of the sandpile. But it leaves untouched the original question of why the central particle in the base doesn't experience the greatest pressure.

One of the principal difficulties in making our worlds of symbols and rules match up with the real world of observations and facts is the eternally slippery nature of language, mathematical or otherwise. The philosopher Ludwig Wittgenstein was tormented throughout his life by the problem of how to describe the nature of the relationship between words and the objects they represent. Wittgenstein is almost unique among philosophers in having created two complete schools of philosophical thought during his lifetime, the second of which completely repudiates the first! Nevertheless, it's the first of these schools, the so-called picture theory of language, that concerns us here.

In his picture theory, Wittgenstein regarded the statements making up a language as being analogous to a series of pictures. Moreover, since he assumed that the logical structure of language mirrors the logical structure of the real world, it followed that these language pictures represent possible states of the world. The theory contends that linguistic statements are meaningful only when they can be correlated with actual states of affairs in the world. Observation then tells us if the statements are true or false. For example, the statement "Roses are red and bloom in the spring" is meaningful since it corresponds to an actual state of the world. But the statement "Bloom roses are the in red and spring" is meaningless within the logical rules (grammar) of ordinary English since it does not correspond to any observable state of affairs in the real world.

The main claim of Wittgenstein's picture theory is that there must be a link between the logical structure of a given language and the logical structure of a real-world fact that a statement in that language asserts. Since this link is itself a relationship in the real world, it's reasonable to suppose that there is some way to express the character and properties of the link using the grammatical rules of the language. But after years of struggling with exactly how to do this, Wittgenstein came to the conclusion that the link between the real world and its expression in language cannot really be "said" at all using language; rather, it can only be "shown." We can't express everything about language using language itself; somehow we must transcend the boundaries of language. Thus Wittgenstein says that we cannot really speak about the world, but only "point." This idea is well summarized by the following statement from the penultimate section of the *Tractatus Logico-Philosophicus*, the only work of Wittgenstein's published during his lifetime:

> My propositions are elucidatory in this way: he who understands me finally recognizes them as senseless, when he has climbed out through them, on them, over them. (He must so to speak throw away the ladder, after he has climbed up on it.)
>
> He must surmount these propositions; then he sees the world rightly.

To illustrate this perhaps disheartening result, consider the relationship between a mathematical model of an economy and a set of real-world observations (measurements) of that economy, things like unemployment levels, interest rates and money supply. Our first step is to translate these observations into some language. So, for example, we might introduce the symbols *U*, *I* and *M* to represent these empirically observable quantities. If we assume that these symbols are words in the language of mathematics, then there is a grammar specifying how we are allowed to combine them in various ways. For example, we can write down the sentence $U + I = M$, which then serves to model mathematically the following dubious, but grammatically correct, relationship: "The sum of the unemployment rate and the interest rate equals the money supply." And, if we like, these symbols can be encoded into the symbols of a computer-programming language, thereby translating the model from the language of mathematics to the language of computing machines. What is the relationship between the logical structure of these symbols within the world of mathematics and the logical structure of the real-world economic quantities the symbols represent? This is what we might without exaggeration term the Fundamental Question of System Modeling.

We have assumed that there is a connection (i.e., a logical structure) linking the real-world observables unemployment, interest rates and money supply. We have also assumed there is a grammatical linkage between the symbols U, I and M in the mathematical world (the grammatically correct mathematical expression $U + I = M$). The degree to which this model reflects reality then depends entirely on how the logical structure of the model and the logical structure of the real-world observables match up. And it is exactly this relationship that Wittgenstein claims cannot be expressed in language (in this case, the language of mathematics). Instead it can only be shown, perhaps by pointing out specific instances of unemployment, interest rates and money supply.

What's important to note here is that the very thing we most care about—the nature of the relationship between what we see and our linguistic description of that observation—is exactly what Wittgenstein says we cannot express in language. And this conclusion holds regardless of the language used to compose the description, including the scientific languages of mathematics and computer programs. This point is of the greatest importance when it comes to distinguishing what's simple from what's not, both in nature and in life, since it forms the basis for the unbridgeable gap between the real world and our models of that world. So without exaggeration we can state that in one way or another it's the limits of language that serve as the deep reason for the emergence of surprises. This theme will be our leitmotif as we go from topic to topic and example to example throughout the book. Now let's look at the closely related issue of how the description we use to characterize a real-world object like a word or a number or a picture colors how complex we think that object might be.

Suppose I give you the two twenty-letter "words"

ABABABABABABABABABAB and QZYRMWERLURLTYCNPBNE.

If I ask which of the two you think is simpler, chances are pretty good you'll say the first word looks simpler since it has a readily recognizable pattern, namely, a boring repetition of the pair AB. Thus, we can compactly characterize the first word by the description "Repeat AB ten times." And, in fact, essentially this same description would work equally well even if the word were 20 million or 20 zillion letters long. By way of contrast, the most compact characterization of the second word seems to be just the word itself. The cartoon in Figure 1.2 illustrates this point.

This elementary observation forms the basis for most characterizations of the complexity of an object: The complexity is directly proportional to the length of the shortest possible description of that object. As a corollary, we can give a rather clear-cut condition for something to be random (i.e., maximally complex): A string of letters is random if there is no rule for generating it whose statement is appreciably shorter—that is, requires fewer letters to write down—than the string itself. So an object or pattern is random if its shortest possible description is the object itself. Another way of expressing this is to say that something is random if it is *incompressible*. This idea forms the basis for the theory of algorithmic complexity, of which we'll have much more to say later in the book.

The idea of using the length of the most compact description to characterize complexity has far broader currency than just as a way of speaking about the complexity of a string of letters. After all, words are strings composed of symbols like, A, B, . . . , Z, just as numbers are strings made up of the symbols 0, 1, . . . , 9. Consequently, the notion of economy of description allows us to talk about the complexity of patterns of any sort, not just patterns of letters and numbers. We can consider the complexity of pictures, sculptures, symphonies and just about any other type of object formed by putting symbols together in particular ways. And in all of these cases the line of

Cotsworth here claims to have found a simpler version

FIGURE 1.2. THE SIMPLE AND THE COMPLICATED

demarcation between the simple and the complicated is a fuzzy one, depending on how hard it is to communicate the pattern of symbols to someone else using a given language. This is clearly a subjective matter, influenced by the richness of the chosen language, the cleverness of the person doing the describing and the listener's ability to use and understand the language. So we see that the subjective everyday notion of complexity is really more a property of the *interaction* between two systems than it is an intrinsic aspect of a system taken in isolation.

Observations and real-world facts are the building blocks from which we construct our visions of reality. But each of these views is merely a small slice of reality, basically a piece in a cosmic jigsaw puzzle. It's the process of putting the pieces together to form ever more accurate pictures that constitutes what in science we call model building. And it's these models—the abstract pictures of reality—that we use to predict future states of the world from the past and present. The difference between these projections and what actually happens then forms the basis of what we colloquially label surprises. But since the notion of surprise is a bit like other informal—yet very useful—notions like truth, beauty, charm, good and evil, in order to sketch the outline of a science of surprise we need to formalize the informal, so to speak. And this process of formalization leads directly to the consideration of just what it is that characterizes a *scientific* approach to the study of surprise, life, the universe or anything else.

RULES OF REALITY

In late 1989, the investment house of Drexel Burnham Lambert made the following forecast: "There are several alternative outcomes which are possible for the U.S. economy in 1990," and went on to list various possibilities ranging from a recession to an inflationary boom. It's predictions like this that give economics and economists a bad name, preventing the field from being taken seriously as a science even by many of its own practitioners.

But since prediction, along with its kissing cousin explanation, is one of the twin pillars upon which the scientific enterprise rests, it's of more than passing concern in our study of the surprising and the complex to ask what it would take to be able to say with a straight face that economics or any other field of intellectual activity is in any way scientific, in either its methods or its results. This question transcends mere academic interest. In today's world a large number of political decisions on everything from environmental pollution to abortion hinge on so-called scientific evidence and the testimony of scientists. So with science being held up as the standard against which all other reality-generating mechanisms are to be judged, it's important for every concerned citizen to have some understanding of just what does and doesn't constitute a scientific answer to a question. And it's of equal importance to understand the kinds of problems that the methods of science are good—and not good—at solving.

At its heart, science is concerned with the question "Why do we see what we do and not see something else?" The scientific answer to this question takes the form of a set of rules, essentially a computer program, by which we can explain what has been observed and/or predict what will be seen next. But rules come in many flavors, and certainly not all of them qualify as scientific. For example, the Ten Commandments are certainly a set of rules. Moreover, they help explain the empirical fact that the majority of people are not regularly engaged in robbery, murder or other extreme types of antisocial mayhem. But hardly anyone would consider these rules to be in any way scientific. Similarly, the astrologer's rule "Saturn in the Second House predisposes one to financial misfortunes" is also considered unscientific today, although in an earlier era it might well have been regarded as the height of scientific respectability. This shows, incidentally, that what is and is not scientific is a time-dependent phenomenon, and that scientific rules—or "laws"—are not as absolute as a lot of scientists would like to believe. But if the Ten Commandments and "Saturn in the Second House" are not scientific rules for predicting and/or explaining observed phenomena, what is?

Basically, there are two quite different sets of criteria that must both be satisfied for a rule to have even a chance of being scientific. The first pertains to properties of the rule itself, while the second has to do with the way the rule is arrived at. In regard to the first type of criterion, here is a checklist of characteristics that tend to separate the scientific rules from the pretenders.

- *Explicit*—Scientific rules are explicit, in the sense that there is no ambiguity in the statement of the rule and it requires no private interpretation to employ the rule for prediction or explanation. For example, Newton's laws of motion state an explicit relationship linking the positions, masses and velocities of a collection of material particles. And as long as you understand what is meant by the terms *position, velocity* and *mass*, there is no ambiguity about either what the rule says or how it's to be applied.
- *Public*—Scientific rules are open to public scrutiny. They are presented in the open literature and can be tested by anyone who has the time, money, equipment and desire to do so. The contrast here with other types of rules is clear, especially those arising from many religions, where rules accessible only to the "divinely inspired" often play a central role in forming the tenets of a particular system of beliefs.
- *Reliable*—Scientific rules have stood the test of time. Before they are accepted as legitimate laws of nature by the scientific community, the rules must have succeeded in predicting and/or explaining a

variety of phenomena over a substantial period of time. Of course, this doesn't mean the rules are infallible and cannot be overthrown in the light of new evidence. But generally speaking, the weight of evidence in favor of the rule must be quite overwhelming before we dignify the rule by labeling it *scientific*.

• *Objective*—Scientific rules are objective in that they are relatively free of investigator bias. In other words, the rule is independent of the social position, financial status or cultural background of the investigator. For example, the exponent in Newton's inverse-square law of gravitation is 2 and not 2.315 or $\sqrt{7}$ or any other number besides 2. And this remains the case for any investigator, regardless of that investigator's professional situation, political leanings or bank balance. In short, the rule is observer-invariant.

I hasten to point out that this does *not* mean that different investigators might not formulate the rule in different terms. But all these formulations must eventually turn out to be equivalent if the rule is to be taken seriously as a scientific rule. For example, in the early days of quantum mechanics there were three seemingly different formulations of quantum phenomena, by Heisenberg, Schrödinger and Dirac. Yet upon further investigation it turned out that all three were essentially the same formulation and could be transformed, one to the other, via routine mathematical operations. So what on the surface looked like different rules ended up being the same rule dressed up in different mathematical clothing. This is the kind of objectivity that's characteristic of scientific rules.

So explicitness, public availability, reliability and objectivity are four key properties of rules that tend to separate scientific prescriptions from those that aren't. But earlier we said that these properties constitute just one of two sets of criteria that a rule must satisfy before we label it *scientific*. The other involves the procedure, or method, by which we generate the rule.

If there's even one thing that students in courses on the philosophy of science remember years later, it's the idea of the *scientific method*. This is the process by which many philosophers of science claim that science distinguishes itself from other reality-generation schemes. And it is this process that gives rise to the kinds of rules that serve as candidates for the coveted accolade *scientific*. The principal steps in this process are shown schematically in Figure 1.3. Here we see three

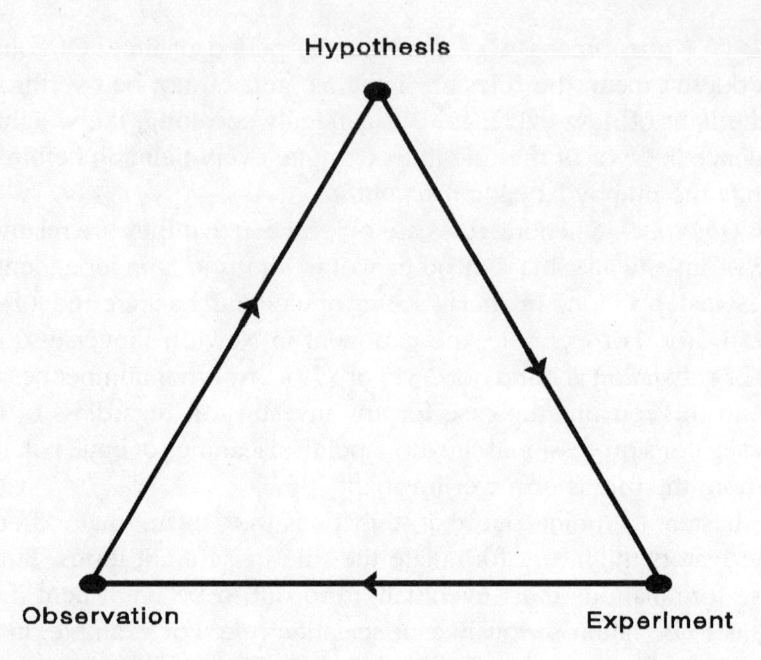

FIGURE 1.3. THE SCIENTIFIC METHOD

stages: Observation, Hypothesis and Experiment. Traditionally, it's argued that the process starts with Observation, and strictly speaking, I suppose this is indeed always the case. But in a field that's already well developed—particle physics, for example—the diagram may just as often be entered at the Hypothesis stage as at the level of Observation. In any case, after a few tours around the diagram we can hope that the process will converge to something. And that something is what serves as a candidate for a scientific rule. Furthermore, if our use of the scientific method does not converge to something, we generally give up trying to fit the observed phenomena into the framework of science.

Putting together the foregoing remarks and observations, we come to the conclusion that the term *science* is more a verb than a noun. Science is something that people *do*, not a property that distinguishes one field of intellectual endeavor from another. In short, science is one particular way of picturing the world. And the scientific way to create these pictures of reality is to produce a "good" set of rules (i.e., a mathematical model) by which we can predict and/or explain phenomena of concern. So from this perspective it makes no sense to claim that, say, physics is scientific while, for example, sociology is not. Either may or may not be scientific. It all depends on whether the fundamental questions in the field are answered by producing a set of scientific rules. So as we wend our way through the thickets of surprising behaviors sprinkled throughout this volume, the reader should keep this point uppermost in mind when considering whether we have yet succeeded in creating a *science* of surprise.

PATTERNS, PUZZLES AND PARADOXES

In 1989, British mathematical physicist Roger Penrose published *The Emperor's New Mind*, a book in which he advances a variety of tantalizing speculations linking quantum theory, the brain and the mind. Penrose concludes that the workings of the human mind cannot be duplicated by a machine—even in principle. This is a comforting conclusion for many, a fact that no doubt contributed mightily to the appearance of such a technical book on the best-seller list for several months. But as thought-provoking as these ideas are, my own guess is that history will be far kinder to Penrose for quite another intellectual achievement, one far removed from rarefied science-cum-philosophical speculations about minds, men and machines. This down-to-earth accomplishment involves the seemingly mundane problem of covering a flat surface like a floor with tiles. It's but a small step from the consideration of such tiling patterns to the problem of compact descriptions considered above. Let me illustrate the situation with an example from crystallography.

Common table salt is formed from atoms of sodium and chlorine arranged in the simple crystalline array shown in Figure 1.4. If we shine a light (or, more properly, a burst of X rays) through this array, the atoms will cast a shadow, or diffraction pattern, like that shown in the left half of Figure 1.5.

If you tried to explain the salt pattern to a colleague over the telephone, you'd have no trouble giving a very short, yet complete, description of it. Maybe you'd say something like "An equally spaced rectangular grid of alternating chlorine and sodium atoms." By way of contrast, the right half of Figure 1.5 shows a the diffraction pattern of what's called a *quasicrystal*. While there certainly appears to be some sort of structure here, it's a pattern that seems difficult to

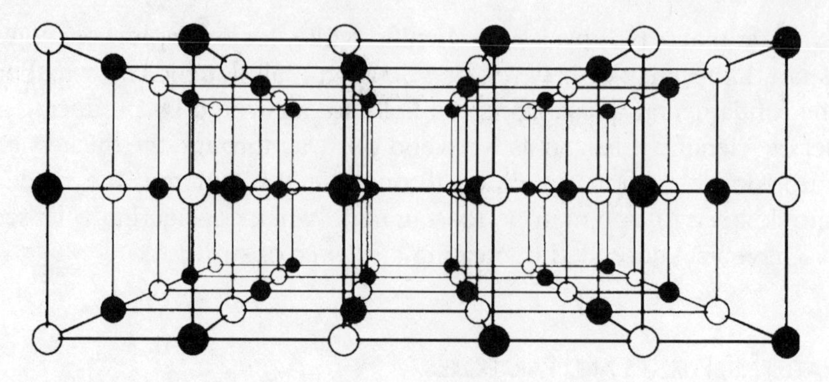

FIGURE 1.4. THE CRYSTALLINE STRUCTURE OF SALT

characterize in any simple, compact way. In fact, these quasicrystal diffraction patterns arise directly from the projections of what's called a *Penrose tiling* of the plane. Such a tiling, in which a planar surface like a bathroom floor is completely filled by a collection of "darts" and "kites," is shown in Figure 1.6.

The point of this example is again to show that when it comes to distinguishing a simple pattern from one that's complicated, a good criterion is to consider the shortest possible description we can give for each of them. This agrees with the intuitive, commonsense view that the shorter the description, the simpler the pattern. As an aside, it's worth mentioning that the relative complexity of the Penrose tiling

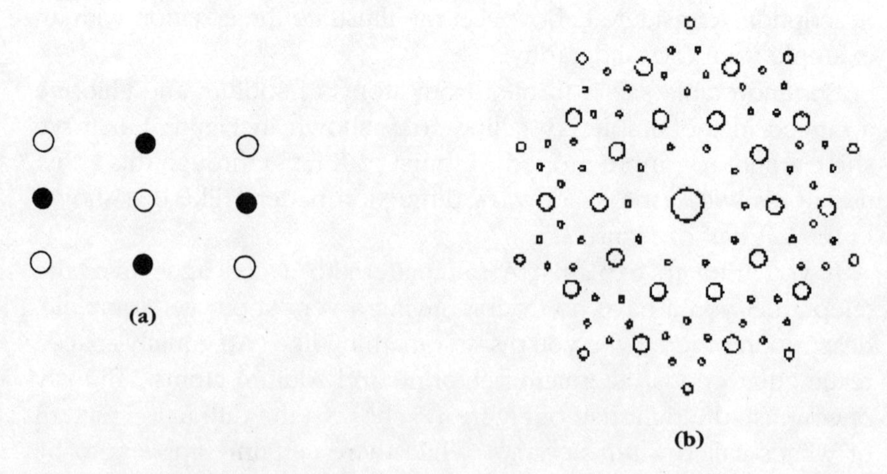

FIGURE 1.5. DIFFRACTION PATTERN FOR (A) SALT AND (B) A QUASICRYSTAL

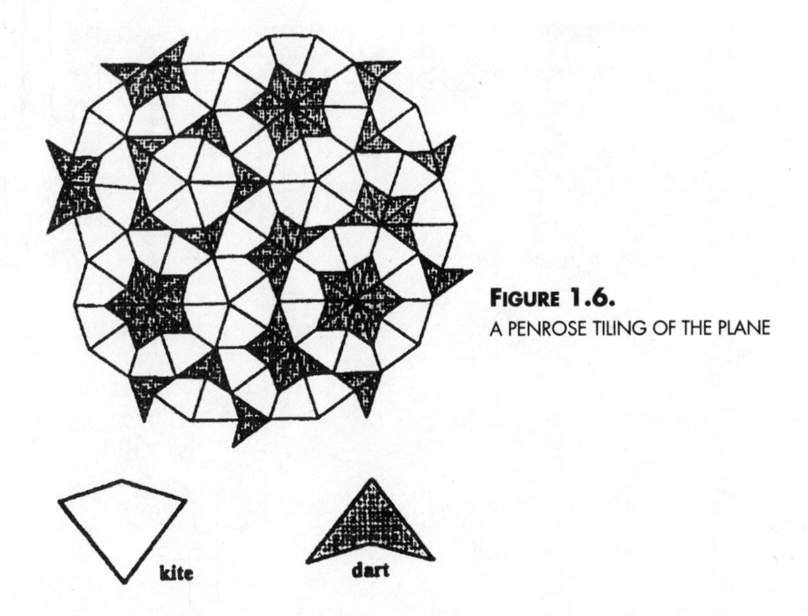

FIGURE 1.6.
A PENROSE TILING OF THE PLANE

leads to many fascinating applications in geometry as well as in the physics of quasicrystals. Moreover, tile manufacturing firms have discovered that the entrancing beauty of the pattern holds considerable commercial appeal, as well.

By way of playing with this notion of using the length of the shortest description as a measure of complexity, the reader might enjoy trying to pin down the complexity of the painting *Checkerboard, Bright Colors 1919* by the Dutch artist Piet Mondrian, shown in Figure 1.7. On the one hand, we have the regularity of Mondrian's checkerboard grid, which argues for low complexity (easy to describe). On the other hand, the colors appear in a seemingly haphazard fashion (hard to describe).

Not only are patterns sometimes difficult to describe, as with the Penrose tiling above, they can also be rather difficult to recognize. This is the essence of those multicolored dot patterns used by eye doctors and driver's license bureaus to test for color blindness. A specific pattern (some number, usually) is hidden within a snowstorm of various colored dots so that it will be invisible if you happen to be blind to certain colors.

A universal feature of knowledge is that one must get outside of a system in order to really understand it. For example, it's just not

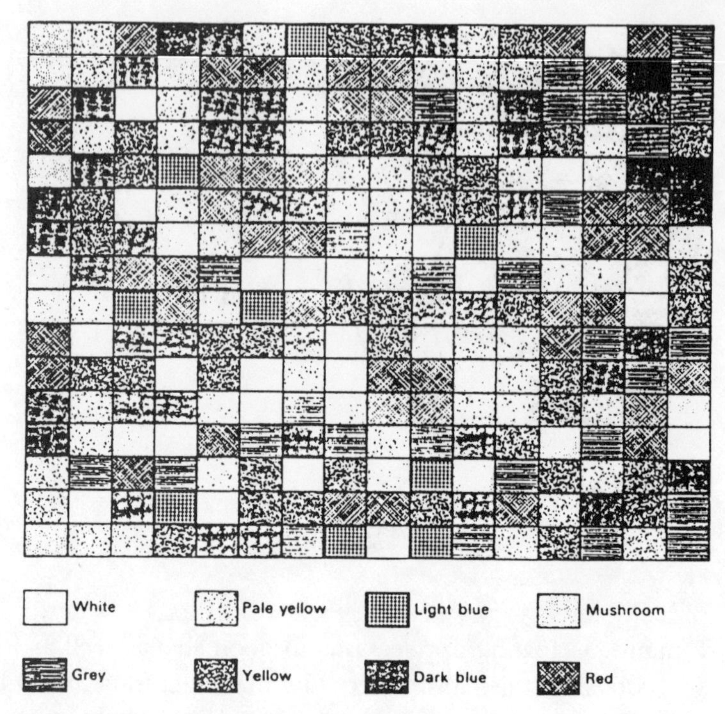

FIGURE 1.7. MONDRIAN'S *CHECKERBOARD, BRIGHT COLORS 1919*

possible to understand the essential nature of a submarine without watching it move beneath the surface of the ocean. Clearly, you can't do this from inside the submarine itself. And as we'll see in a later chapter, Gödel's Theorem in mathematical logic tells us that the truth or falsity of statements in a branch of mathematics cannot be determined without going beyond that branch of mathematics to another. So one way of dispelling confusion in pattern recognition tasks is to "jump out of the system," so to speak, and look at the problem from a different hierarchical point of view. Let me illustrate this kind of hierarchical leap with an episode from my own mercifully brief career as a graduate student.

Before enrolling for graduate studies at the University of Southern California in the late 1960s, I had heard rumors to the effect that USC students and fans take their sports seriously, especially football. While applying the finishing touches to a doctorate in mathematics in the fall of 1969, I was inadvertently caught up in this football mania one sunny Saturday afternoon at the Los Angeles Coliseum. On arriving at the Coliseum and taking my assigned seat in the USC student rooting section, I was surprised to find a large manila envelope awaiting me. When I opened the envelope, out spilled several large cards in assorted colors, together with a small instruction sheet congratulating me on being elected to a one-game membership in the famed USC flashcard section. The instructions went on to say that my job was to listen carefully to the cheerleader during the half-time show, and hold up my card numbered *X* when the cheerleader called out stunt number *Y*. Under the wildly optimistic assumption that everyone in the card section would follow their assigned task, the cards held up would then be visible to the TV cameras across the way as a particular pattern—probably something showing Tommy Trojan soundly thrashing a Cardinal or a Bruin or, even better, a Fighting Irishman. But the one group of people in the Coliseum that day who would never know which of these patterns was actually displayed was my group—the "flashers."

That's the point. In order to see the pattern formed by the cards, you have to jump outside the system. Down at the level of the system itself (the level of the individual flashers like me), there was no recognizable pattern but only a seemingly senseless holding up of one colored card or another. The reader will recognize the similarity of this situation to that of a Times Square message board, in which individual lights flash on and off to form a pattern (i.e., a message) that can only be seen and understood at a level beyond that of the lights themselves. Here we have a situation in which common sense living at one level might well argue that there's no pattern of any sort to be seen. Yet by moving outside the system to a different view of the situation, pattern and structure emerge as if by magic.

But common sense can be fooled in many other ways, too. One of the most entertaining is by paradox, an enemy of clear thinking that comes in a bewildering variety of forms—artistic, logical, linguistic and otherwise. Let's look at a few examples of the ways paradox can generate surprises by leading common sense astray.

- *The Alabama Paradox*—Following the census of 1880, the seats in the U.S. House of Representatives had to be reapportioned to reflect the new population distribution. The apportionment scheme in use at the time had the totally unexpected property that when it was applied to a House with 299 seats, the state of Alabama was entitled to 8 seats, but when the size of the House was set at 300 seats, Alabama's representation was actually *reduced* to 7 seats.

Needless to say, the loss of a Congressional representative caused considerable furor at the time (especially in Alabama), along with the call for an explanation of such a counterintuitive outcome. After all, if the total number of representatives is increased, how could any state's fair share of seats go down? This is the *Alabama paradox*. It was not until more than thirty years later that the reapportionment scheme that's still used today was introduced, a scheme that explicitly precludes a state's losing seats if the total number of seats in the House is increased.

It's not very important for our present purposes why the old apportionment procedure led to such a counterintuitive result. We'll see some of the reasons in a later chapter. Here it's only necessary to note that what looks like a perfectly sensible, fairminded set of rules for dividing up a resource can lead to a result that flies straight in the face of intuition and common sense. But reason and logic are not the only things that can be fooled by paradox. Here's another example. This time it's our eye that's fooled instead of our mind.

- *The Impossible Staircase*—In a 1958 issue of the *British Journal of Psychology*, the geneticist L. S. Penrose and his son Roger, creator of the Penrose tiling discussed earlier, published the visual illusion now known as the *impossible staircase*. It's shown below in Figure 1.8(a). What's paradoxical about this staircase, of course, is that as you make a complete tour of the stairs and come back to where you started, with each step you appear to be moving to a higher level, yet you come back to exactly the same level at which you began. This illusion of continually rising to higher and higher levels was later put to good use by the Dutch artist M. C. Escher in many of his most famous engravings. A good illustration is his well-known work *Ascending and Descending*, shown in Figure 1.9.

The secret of the impossible staircase can be unraveled by slicing it as shown in Figure 1.8(b). Here we see that if you follow level 1 from its highest apparent position at the upper right, it reappears at the base of the staircase structure. Similarly, level 2 also reappears at the bottom of the structure, just above level 1.

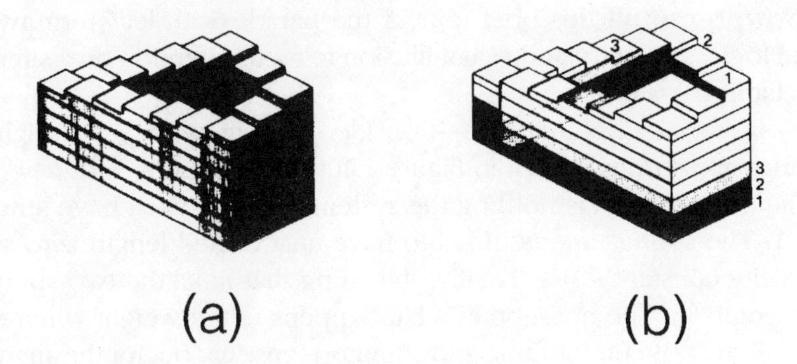

FIGURE 1.8. THE IMPOSSIBLE STAIRCASE

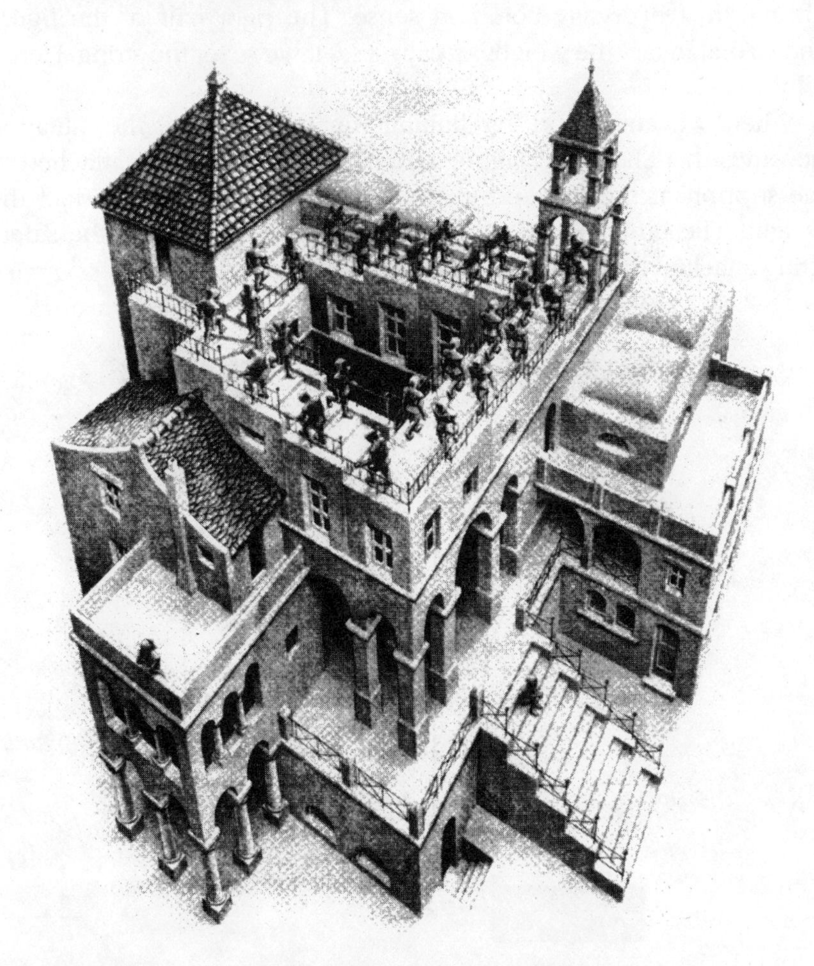

FIGURE 1.9. M. C. ESCHER, *ASCENDING AND DESCENDING*

Now to cap off this brief tour of the paradoxical, let's turn away from logic, geometry and visual illusion to focus attention on a simple mechanical system.

• *A Mechanical Network*—Consider the network of weights, springs and strings shown in Figure 1.10(a). Here the weight $w = \frac{1}{2}$, while the two strings holding the system together each have length $L = 1$. The springs are assumed to have unstretched length zero and elasticity constant $k = 1$. Finally, the string that links the two springs has length $\frac{3}{8}$. The question is What happens to the weight when the linking string is cut? Intuition and common sense cry out for the answer that the weight drops. But by now we should know that you just can't always trust everyday common sense! The right half of the figure shows that in fact the weight actually *rises* if we sever the string. Here's why.

When we cut the string linking the two springs, the situation becomes that shown in Figure 1.10(b). The safety string attached to the support at the top and to the lower spring now bears half the weight. The other half is borne by the upper spring and the safety string attached to it. Thus, the extension of each spring is $\frac{1}{2} \times \frac{1}{2} = \frac{1}{4}$.

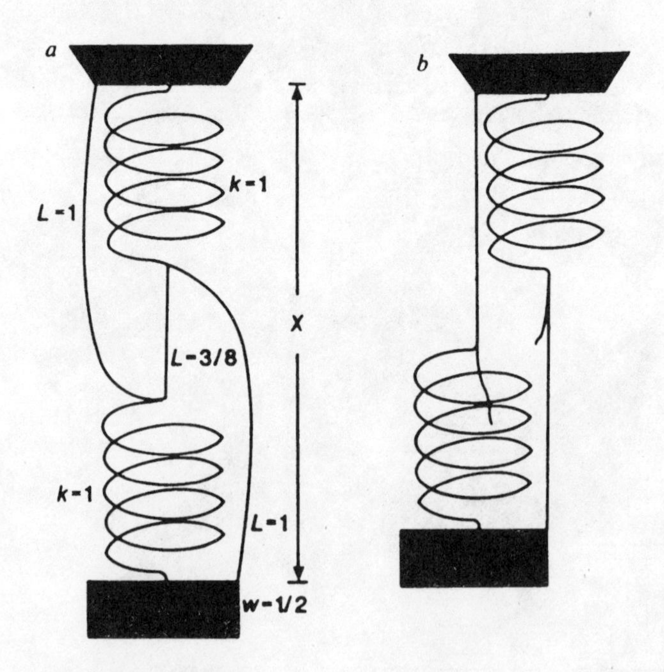

FIGURE 1.10. WEIGHTS, SPRINGS AND STRINGS

As a result, the distance from the support to the weight is now $1 + \frac{1}{4} = 1\frac{1}{4}$—less than its previous length of $1\frac{3}{8}$. Consequently, at equilibrium the weight must be higher than its original position. But not even counterintuitive networks can violate the laws of physics, and the potential energy of the system is now less because the springs have contracted from their previously extended lengths. This reduction in potential energy of the springs compensates for the energy gained by raising the weight.

In case you're thinking that this kind of network is merely a theoretical curiosity devoid of any practical interest, the situation depicted in Figure 1.11 will quickly sober you up. It shows how just such a paradoxical rearrangement of nuts and bolts resulted in a considerable loss of life a few years ago in a Kansas City hotel. As shown, a "routine" engineering change from (a) to (b) during construction of the hotel tripled the shear load on nut N_1, causing three hotel balconies to fall when the nut split open. The paradox here is that configuration (b) distributes the load over seven nuts instead of three as in (a). Nevertheless, the series-to-parallel transformation underlying these kinds of network paradoxes ends up placing a greater load on nut N_1 when it's in configuration (b) than in (a).

This simple mechanical paradox has analogs in electrical circuit theory, traffic flow and hydraulics as well, illustrating the possibility of counterintuitive behavior in many types of networks when the load imposed on an arc in the network affects that arc's behavior (in this case, the stretching of the spring). These sorts of paradoxes themselves constitute surprises arising out of a mismatch between what common

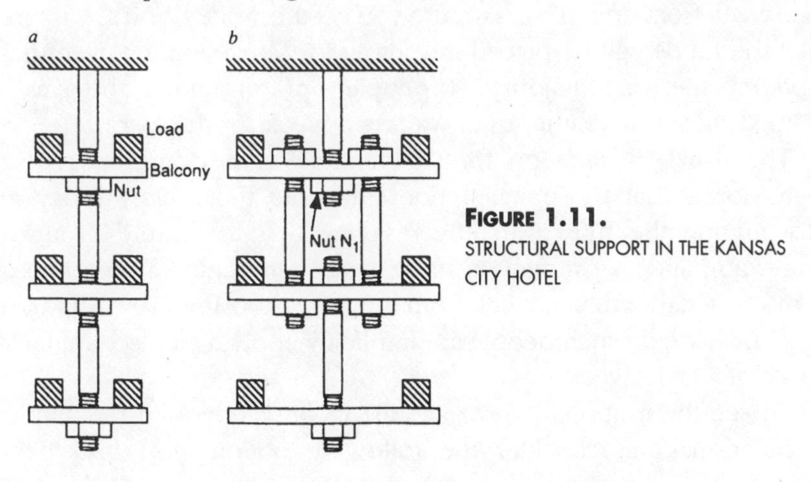

FIGURE 1.11.
STRUCTURAL SUPPORT IN THE KANSAS CITY HOTEL

sense and intuition (i.e., our models) tell us and the way the world just happens to be constructed.

Now let's turn from simple geometrical patterns and paradoxes to a particularly widespread form of cognitive contradiction—the phenomenon of irrationality.

Common Sense and Irrationality

The refinement in economic thought over the past two hundred years would almost certainly make today's theories of men and money unrecognizable to Adam Smith. Yet one point upon which the Adam Smith of 19th-century Scotland and the Adam Smith of the PBS television show "Money Line" would certainly agree is that human economic behavior is ultimately rational. Economists have always believed that individuals make choices so as to maximize their *utility*, meaning that we try to derive as much satisfaction as possible from the sum total of our actions. So we watch football on TV until we think our hours and minutes can be more happily spent listening to Mozart or reading *War and Peace*. And we spend our meager paychecks on beans and bananas unless cream cheese and caviar seem to hold greater allure. Utility maximization, then, is a form of rational self-interest. But despite its elegance as a theory, utility maximization just doesn't seem to match up to the way the real world actually works. Let's look at an example given by Richard Herrnstein and James Mazur showing how the commonsense, utility-maximizing solution to a problem can actually be profoundly irrational.

Suppose you are given the choice of receiving $100 today or $120 one week from now. To make sure you get the money, you are assured that the funds will be placed in escrow for safekeeping. Given these options, the vast majority of people opt for taking the smaller immediate reward rather than waiting a week for the extra $20.

The usual explanation for this take-the-money-and-run type of behavior is that those who choose to take the money today are discounting the future, in effect saying that the future is always uncertain, and that money in hand today can begin earning interest tomorrow rather than a week from now. By choosing $100 today over $120 next week, such people are implicitly applying a discount factor of 20 percent a week.

To see the irrationality—or at least inconsistency—in this kind of decision-making, consider the following option: you may either receive $100 today or wait one year and get $1,300,000. Just about everyone I know answers that they would be perfectly willing to wait a year for that million bucks, as opposed to taking a quick hundred today. Yet these are the same 20-percent discounters who said that $100 in the hand is preferable to $120 in the bush. If they were consistent (i.e., rational) decision-makers, they would still take the immediate payoff of $100, since the larger amount discounted at 20 percent a week for a year amounts to only $99.20. So utility maximization suggests indifference between the two choices, at best. But real people are far from indifferent to these two alternatives, expressing a pronounced preference for the "big score."

When confronted with this sort of experimental evidence showing that real decision-makers act irrationally in that they do not maximize their utility, economists and psychologists often explain the apparent breakdown of rationality by appealing to the myriad factors involved in the decision-making situation, things like incomplete information, too short a decision-making time and limited data-processing capabilities. But as we'll see in a later chapter, these factors are not the main culprits at all. Rather, the root cause of the difficulty lies in the relationship between the average utility over time and instantaneous, or *marginal*, utility. So to resolve the irrationality dilemma, we have to appeal to a theory of *approximately* rational behavior, or what psychologists term *melioration*.

So far we have concentrated on the kind of complexity seen in static structures—artistic patterns, logical puzzles, one-time-only decisions and the like. "But," as Galileo once said, "it moves." And so we're led inexorably to consider dynamical processes, where the most difficult-to-understand behaviors can unfold from what seem like the most humble of origins. Our next section takes a quick look at some of the strange things that can happen in the wonderland of dynamical processes.

IT'S ALL IN THE MOTION

Treasure Hunt is a popular childhood game, in which each player receives a list of clues that point to the contest prize. One common variant involves having the original list of clues lead to a second list, which in turn leads to a third list and so on, the final list of clues then pointing to the actual prize itself. The path followed by any one player in this version of Treasure Hunt serves admirably as an example of a dynamical system, the crucial ingredient needed to understand the appearance and disappearance of complicated patterns and behaviors over the course of time. So let's follow the fortunes of a single player in this kind of Treasure Hunt, calling our man John.

Assume for the sake of definiteness that the playing arena for the hunt consists of the entire territory of Central Park in New York City. The game then begins with John receiving the initial set of clues while standing at, say, the Plaza Hotel entrance to the park, which we might conveniently term "ground zero." Decoding the clues on this initial list, John moves from the Plaza entrance to the foot of a large oak tree next to the Zoo, where upon moving a few leaves and branches aside, he finds the second set of clues. Musing a bit on these new clues, John soon teases out their meaning and takes a hike to the opposite side of the park over by the Lake. Digging a few inches into the sand by the shore of the Lake, John uncovers the third set of clues. The message coded into these clues tells John to go to the delivery entrance behind the Tavern On The Green, where the "treasure," a coupon worth dinner for two at any restaurant in the city, lies hidden beneath the third garbage can on the left. Figure 1.12 gives a pictorial account of John's tour in moving from his starting point at the Plaza to the ultimate promise of a fine meal. For later reference, the reader should note that Figure 1.12 also shows the path taken by Jim, one of John's competitors in the game, who started at the Fifth Avenue entrance to the park rather than at the Plaza Hotel. This process of following the clues from the Plaza entrance to the Tavern On The Green is a concrete example of what scientists call a *dynamical system*.

A dynamical system is composed of two primary ingredients: (1) a playing field, or space, on which the motion of the system takes place, and (2) a rule telling us where to go next from wherever we are now. Technically, the playing space is termed a *manifold,* while the rule of motion is called a *vector field*. So in the case of John and the Treasure Hunt, the manifold is Central Park, a rectangular region bounded on the north by 110th Street, on the south by 59th Street, on the west by Central Park West and on the east by Fifth Avenue. The rule governing John's motion over this manifold is the set of instructions coded into the lists of clues John found at the various locations in the park. This set of three lists constitutes the vector field of this dynamical system.

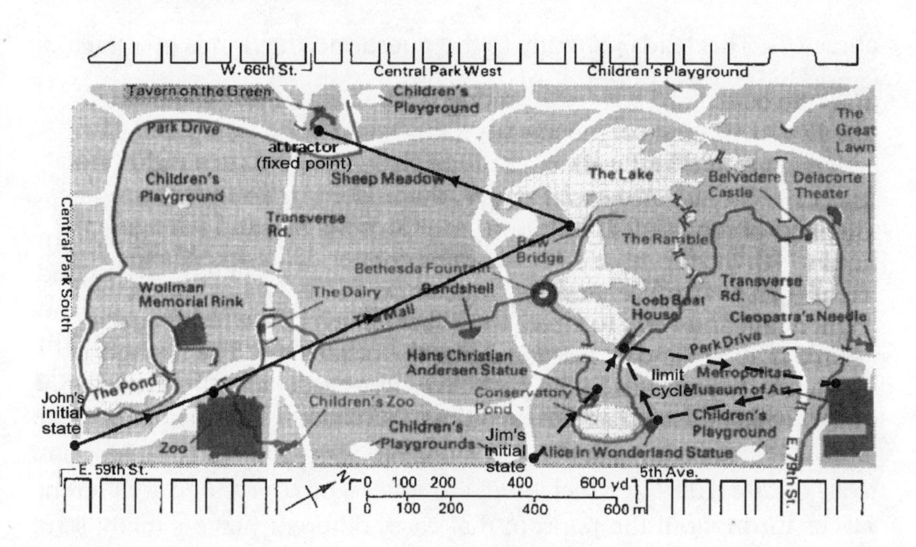

FIGURE 1.12. JOHN'S PATH THROUGH CENTRAL PARK

Finally, the path John traced out as he moved from one location to another is called the *trajectory* of the system, while the starting location at the Plaza entrance is termed the system's *initial state*. It's clear from this setup, I think, that once the manifold, the initial state and the vector field (the set of rules) are given, John's trajectory through the park is then completely fixed—assuming he decodes the clues correctly and that each list of clues defines a unique location in the park.

What's of great interest to both dynamical system theorists and to John and Jim is the end point of the trajectory, what's called the system's *attractor*. Perhaps the best way to envision the attractor is to think of it as a kind of mathematical black hole. Once the system behavior moves onto the attractor, it stays there forever. There are at least two reasons why the attractor is so important. First, in many physical systems the short-term, or *transient*, motion leading from the initial state to the attractor is very rapid. As a result, what we see when we look at the behavior of such systems is what the system is doing when it's already on (or very near) the attractor. But even if the system is not yet on the attractor, its motion is still governed by the nature of the attractor in much the same way that the path of a leaf or a cork floating in the sea is dictated by the size and nature of nearby whirlpools and waves. In short, the overall behavior of a dynamical system is for the most part fixed by the number and character of its attractors. This leads naturally to the question What kinds of attractors can a dynamical system display?

The Three Attractors

Basically, there are three main types of attractors. The simplest is the *fixed point*, exemplified by the single garbage can behind the Tavern On The Green at which John's trajectory through Central Park ended up. In that case, the clues led to one definite point in the park, and when John got there his motion ceased. At such a point, the rule of the vector field says, in effect, "Stop right here. The game is over." Of course a system may have many fixed points of this sort. The particular one you end up at is determined by where you start. This is analogous to a Treasure Hunt in which there are several prizes located at different places throughout the park. In that case, different players might start at different locations, thereby tracing out different trajectories by following the clues from the different starting points. Some of these paths would lead to the same garbage can behind the Tavern On The Green that "captured" John. We would say that these players and John were all in the *domain of attraction* of the garbage-can fixed point. Other players, though, might end up at other locations scattered throughout the park, each of these places also being fixed points of the dynamical system. These points also each have their own domain of attraction (i.e., sets of players, that are "attracted" to them).

The next simplest kind of attractor comes about when the rules of the system cause a given trajectory to repeat itself in a cyclic fashion. In this case, after an initial transient phase the system's trajectory passes through the same sequence of points forever. This kind of periodic orbit is technically termed a *limit cycle*. In Figure 1.12 we see that after a transient stage consisting of a stop at the Hans Christian Andersen statue, Jim's trajectory is "sucked" onto the limit cycle attractor consisting of the Loeb boathouse, the Metropolitan Museum of Art and the Alice in Wonderland statue—a cyclic orbit of period three, since there are three locations constituting the orbit. Limit cycles are often encountered in dynamical processes used to describe things like clocks, economic fluctuations and the human heartbeat. Fixed points and limit cycles are the two classical types of attractors and were almost certainly known to Newton. Their geometric structure is shown diagrammatically in Figure 1.13. But there is a third, decidedly nonclassical, type.

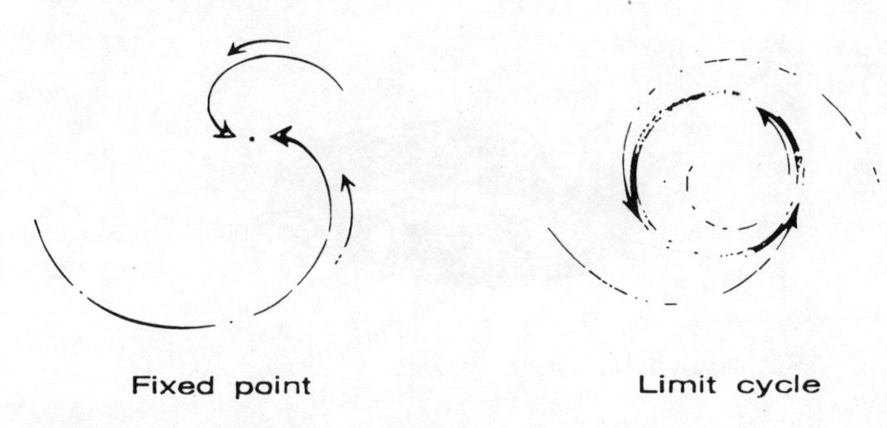

FIGURE 1.13. A FIXED POINT AND A LIMIT CYCLE

The final type of attractor is the most complicated and, surprisingly, the most common. It is what's nowadays called a *strange attractor*. In rough terms, the strange attractor is a lot of periodic orbits and aperiodic paths all rolled up into one big tangled mess. A good visual image of such a monstrosity is a ball of yarn or, even better, a bowlful of spaghetti (with sauce). Each strand of spaghetti in the bowl is one part of the strange attractor, the spaghetti sauce ensuring that no two strands ever quite make direct contact. Yet on each strand there is some point that is as close as we like to any other strand, some strands even closing back on themselves to become periodic orbits. But to be on a strange attractor, periodic orbits must, in general, be of a special type: unstable. This means that the path describing the orbit acts like a repellor rather than an attractor. If we move off such an orbit by even a little bit, we are then pushed even further away to another part of the attractor rather than being attracted back to the original orbit, as would be the case if the orbit were stable. A strange attractor is shown schematically in Figure 1.14. Since these ideas about dynamical systems are so crucial to our subsequent discussion, let's try to cement them in place using a familiar household example.

The magic that makes a grandfather clock work as a timepiece is the regular, cyclical motion of the bob at the end of the clock's pendulum as it swings back and forth between its two extreme positions. Regarding such a clock as a dynamical system, the set of states is the collection of all the positions that the bob can possibly be in, together with the speed and direction of motion of the bob at those locations. Consequently, we can specify the state of the clock

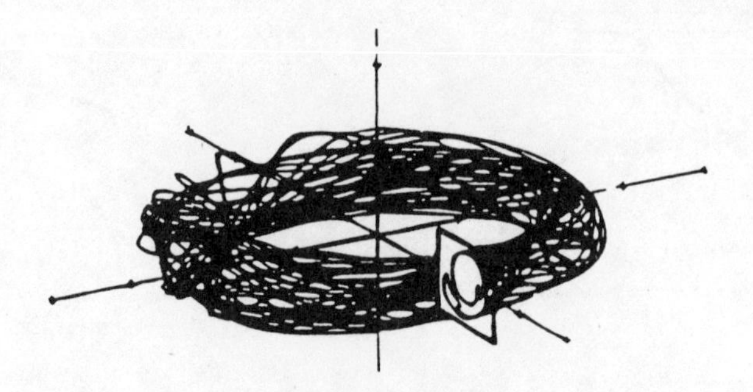

FIGURE 1.14. A STRANGE ATTRACTOR

at any particular moment simply by giving two numbers: the first represents the position of the bob away from its rest point, the second its velocity, say positive for motion to the right and negative if the bob is moving to the left. For the sake of discussion, let's agree to call the state of the clock when the bob is at rest the zero state.

Assume at first that there's no friction in the clock's mechanism (i.e., the pendulum is "undamped"). So when we pull the pendulum bob away from the zero state and let it go, the bob swings to and fro, tracing out the same arc forever. The amplitude of this arc is determined solely by the distance that the bob is pulled away from the rest state. This phenomenon is shown geometrically in Figure 1.15. The picture makes it clear that the attractor for this undamped case is simply a circle in the space of possible states, which are shown in the right half of the figure. In short, the motion is periodic and the attractor is a limit cycle in which there is no transient motion, since the system is already on the attractor when the motion begins.

Now let's forget about idealized frictionless clocks and sprinkle a little dust and sand onto the bearings of our clock's mechanism. Now the clock will run down, with the bob eventually returning to the rest state. Thus, the attractor for this "damped" system is the single fixed point consisting of zero position and velocity, or the zero state. Figure 1.16 shows the way the clock spirals down to this fixed point at the origin in the space of clock states. Now consider the extreme case in which we pull the pendulum bob all the way to the top, so that it stands precariously balanced in a position exactly opposite to that of the zero state. What's the situation with the attractor now? Herein lies a crucial tale in the storybook of dynamical systems.

For normal grandfather clocks whose bobs are not displaced to the

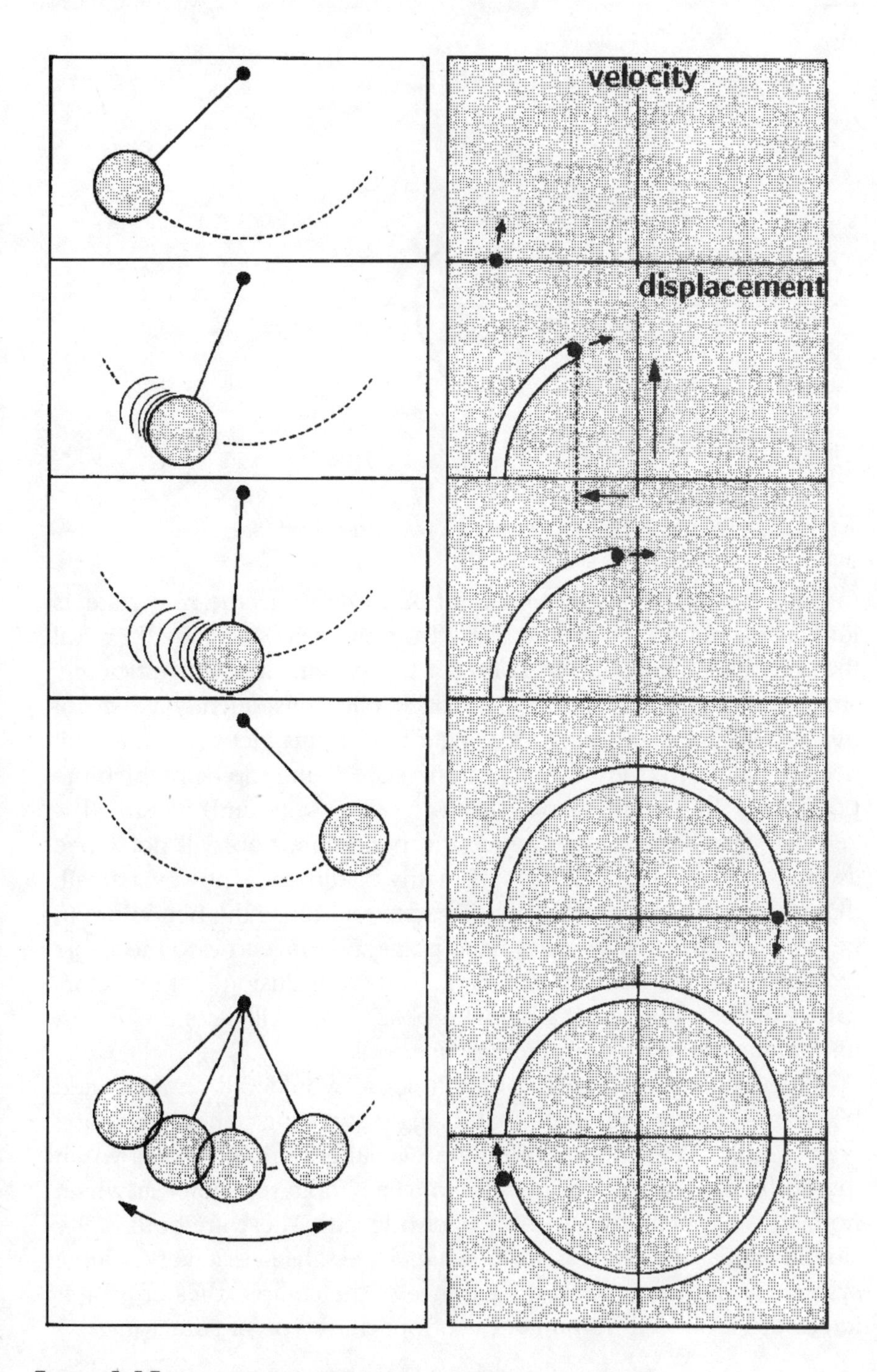

FIGURE 1.15. BEHAVIOR OF A GRANDFATHER CLOCK WITH NO FRICTION

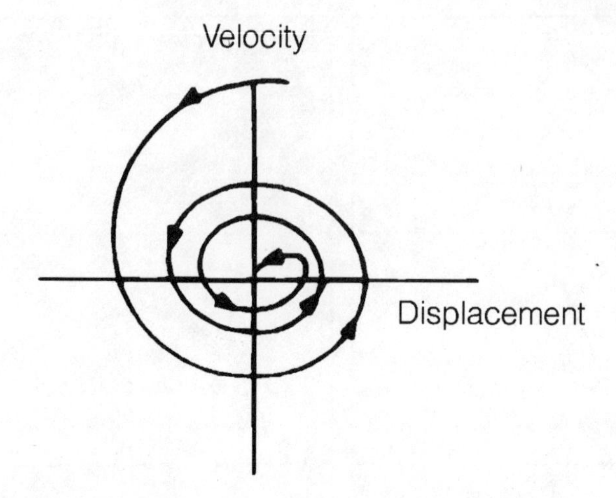

FIGURE 1.16. ATTRACTOR FOR A REAL GRANDFATHER CLOCK

top, the typical behavior is for the bob to return to the zero state as long as it's not pulled too far away from the zero state. We then call the zero state a *stable fixed point* for the system since all sufficiently small perturbations eventually lead back to it. Consequently, we might think of such a fixed point as being a kind of magnet that attracts all nearby states. On the other hand, the state at the top—in which the position is 180 degrees away from the zero state and the speed is zero—is also a fixed point since the pendulum bob will not move away from this state on its own. But any disturbance, however small, that pushes the bob away from this state will cause the pendulum to be repelled from this *unstable* fixed point and move toward the stable fixed point at the bottom. We come to the conclusion that attractors can be either stable or unstable. This observation will be of the utmost importance throughout the rest of the book.

When it comes to what common sense would call "complicated behavior," it's hard to argue that a single swinging pendulum is any way complex. The motion of the bob is either boringly repetitive (in the no-friction case) or winds down to a simple rest point at which nothing is happening (if friction is introduced). These are both pretty simple kinds of behavior from what, in essence, is a very simple system. So in our quest for the essence of the complex, it's of greater interest to look to the strangeness of our third type of attractor.

* * *

In the mid 1970s, Mitchell Feigenbaum was a physicist at Los Alamos in search of a problem. While waiting for inspiration from his muse, Feigenbaum started playing around with a hand calculator, looking at the properties of a series of numbers generated by a very simple scheme involving one addition and two multiplications. What Feigenbaum discovered was one of the sparks touching off the explosion of interest in what we now call chaos, fractals and computer-oriented experimental mathematics. To make a long story short, Feigenbaum's seemingly aimless punching of buttons on his calculator turned up the heretofore unappreciated fact that by following a perfectly deterministic set of simple rules, you can end up with a completely unpredictable, essentially random result. And, in fact, the rules that Feigenbaum used to generate his magic numbers were the very same rules that programmers often employed in the early days of computing to generate a list of numbers giving every appearance of being completely random. To see the import of Feigenbaum's momentous discovery, let's consider an equivalent version of these rules, characterizing what I like to call the Circle-10 system.

Suppose we are given a circle C whose circumference is divided into ten segments of equal length, labeled 0, 1, . . . 9. To make things simple, let's also suppose that the circumference of the circle is 1, so that each section has length $\frac{1}{10}$. With this setup, any number between 0 and 1 can be represented as a point on the circle in the following way: sector 0 corresponds to all numbers between 0 and 0.099999. . . , while sector 1 runs from 0.1 to 0.19999 . . . and so on to sector 9, which contains all the numbers from 0.9 to 0.9999. . . . So, for instance, the number $r = 0.379762341$ is a point lying in sector 3, almost 80 percent of the way to sector 4.

To create the Circle-10 dynamical system, let the points of C be the manifold of states. Consequently, a state of the system is specified by a single number between 0 and 1. Now define the rule of state transition to be "If you're at the point x, go to the point $10x$." But remember that one time around the circle takes you back to where you started, as do two, three or any other whole number of circuits. So, since the circle has a total length 1, when we multiply any number by 10 to get the next point, it's necessary to chop off the integer that appears on the left side of the decimal point. This is the mathematical way of ignoring complete tours around the circle. In other words, to find out where to move next on the circle from a particular point x, we keep only the fractional part of the number $10x$. Figure 1.17 shows the results of the first step of this dynamical process, starting at the point r given earlier, while Table 1.1 gives the first nine steps on the trajectory of the Circle-10 dynamical system from this same initial point r. From this discussion, we see that the action of the Circle-10 rule is the ultimate in simplicity: wherever you are now, just multiply by 10 and delete the digit to the left of the decimal point to get the address of where you should go next. It's hard to think of a rule easier to describe than this.

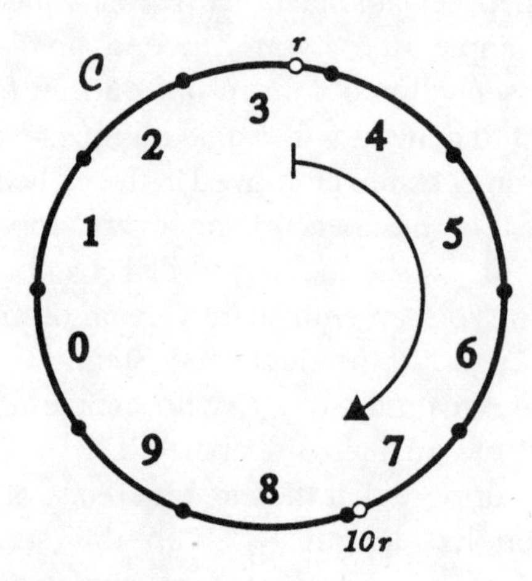

FIGURE 1.17. THE CIRCLE-10 SYSTEM

TABLE 1.1.
THE FIRST NINE STEPS ON THE CIRCLE-10 TRAJECTORY

Time	*Number*	*Point on Circle*	*Sector*
0	0.379762341	= 0.379762341	3
1	3.79762341	= 0.79762341	7
2	7.9762341	= 0.9762341	9
3	9.762341	= 0.762341	7
4	7.62341	= 0.62341	6
5	6.2341	= 0.2341	2
6	2.341	= 0.341	3
7	3.41	= 0.41	4
8	4.1	= 0.1	1
9	1	= 0	0

Now look at the sequence of sectors that the system visits. From Table 1.1 we see that the itinerary of the starting point r is sectors 3,7,9,7,6,2,3,4,1,0,0,0,0,0, ... If these numbers look familiar, they ought to—they are just the decimal digits of the starting point r! And this is not just a coincidence. For *any* starting point, the list of sectors visited exactly matches the decimal digits of the starting point for the simple reason that the rule "Multiply by 10 and chop off the digit to the left of the decimal point" corresponds to nothing more than shifting the decimal point one position to the right and neglecting complete circuits. This is about the most straightforward, easy-to-calculate, deterministic dynamical system imaginable.

But simple to calculate and describe doesn't necessarily mean simple in behavior, and the Circle-10 system captures almost all of the interesting behavioral features associated with chaotic systems and strange attractors. Let's take a look at some of the most important of those features.

- *Divergence*—A number is called a *palindrome* if it reads the same backward and forward. Thus, the palindromes are 1, 2, 3, . . . 9, 11, 22, 33, . . . 99, 101, 111, 121, . . . 191, 202 and so on ad infinitum. Suppose we form a single number in the unit interval by writing down all palindromes in order and then putting a decimal point in front of it. Let's call this number P for palindrome. So we have

$$P = 0.12345678911223344556677889910111121131 \ldots .$$

Now suppose we choose a starting point p for the Circle-10 system whose decimal digits agree with those of the palindrome number P in the first billion digits, but thereafter continue with 66666 . . . forever. Thus, the itinerary of the system starting from P agrees with the itinerary from p for the first billion steps. But thereafter the trajectory from p stays put in sector 6 forever, while that from P goes on about its business, whatever that may be, but which definitely does not involve grinding to a halt in sector 6.

This example shows that two starting points, P and p, closer together than we'll ever be able to measure, wind up following completely different paths. In the dynamical system theorist's dictionary, systems having this kind of divergent behavior are said to be *sensitive to initial conditions*.

One way of thinking about this kind of sensitivity is in terms of information. Initially, we have two starting points that are observationally indistinguishable. As time unfolds, the trajectories emanating from these two points diverge to the extent that we are able to recognize two distinct trajectories. This means that there has been a creation of information, in that by knowing there are two trajectories we also then know that since every starting point gives rise to a unique trajectory, there must have been two different starting points—even though we could not distinguish them at the outset. In short, the system has generated information during the course of its operation. This information generation can be measured by something called the *K-entropy*, about which the reader can find more details in the To Dig Deeper section for Chapter Three.

• *Randomness*—Earlier we noted that almost every real number is random in the sense that there is no way to describe its digits that's shorter than just writing them down, one after the other. But this means that for almost every starting point of the Circle-10 system, a listing of the sectors visited will be indistinguishable from the output of a random-number generator. This experiment shows that randomness need not necessarily come from an indescribable or uncertain rule. A deterministic rule applied to a known starting point can generate an outcome that's every bit as random as the spin of a roulette wheel or the toss of a coin.

• *Instability of itineraries*—Almost all itineraries are random, but some are not. Which ones? Obviously these are the itineraries whose starting points consist of digits that repeat themselves after a finite number of steps. Such points lead to trajectories that repeat themselves over and over again. It's a well-known mathematical fact that the numbers having this kind of periodic pattern in their digits are precisely the rational numbers: numbers like 13, 1/2, 7/23 and 192397/209587, each of which is the ratio of two whole numbers.

In the interval between 0 and 1 there are an infinite number of rational numbers, as well as a much larger infinity of irrational numbers like π, $\sqrt{17}$ and the palindrome number P. It turns out that between any two irrational numbers there is a rational one, although they do not alternate (almost all the numbers are irrational). Consequently, the starting points that lead to periodic itineraries are totally mixed up with the *aperiodic points*—numbers whose decimal expansion continues forever without repetition—that do not lead to such repetitive itineraries. This fact also shows that the periodic points are unstable, since if we perturb them just a little bit to a nearby irrational, the system trajectory takes off on a totally different course, one that no longer repeats itself. As it turns out, there can be at most one stable point, with other starting points being unstable. So regardless of whether you start the Circle-10 sequence at a periodic or an aperiodic point, the itinerary is almost surely unstable in the sense that a small perturbation away from any point on the itinerary sends the system off onto a completely different trajectory. The bowl-of-spaghetti picture of a strange attractor is a good way to visualize how this can happen.

At this juncture in our narrative it's reasonable to ask, Are things like the Circle-10 system just mathematical curiosities, or are they something we can expect to encounter in the real world? A clear answer to this eminently sensible query has emerged over the past decade or so: systems with strange attractors are the rule, not the exception. So if you're dealing with mathematical representations of the real world, then you're dealing with strange attractors or, in the vernacular, chaos. We will see ample evidence for this claim in Chapter Three. By way of introduction to these ideas, let's look at a different dynamical process, one that goes under the rubric of the *logistic law* of growth.

Over the past twenty years or so, Cesaré Marchetti of the International Institute for Applied Systems Analysis in Laxenburg, Austria, has been poring over statistical records describing the growth and decline of hundreds of psychological, social, technological and political phenomena. These efforts have turned up the truly remarkable fact that the dynamics of things like Mozart's musical works, construction of Gothic cathedrals in Europe and the volume of world airline traffic all seem to follow the same simple pattern, which biologists call the *logistic*, or *S-shaped, curve*.

The basic form of the logistic curve is shown in the left half of Figure 1.18, which displays the growth of a bacterial colony over time. The right half of the figure shows the same S-shaped curve plotted in coordinates using the quantity $F/(1 - F)$, where F represents the fraction of the final system "size." So, for example, when the ratio $F/(1 - F)$ equals 1, the system has reached 50 percent of its final size,

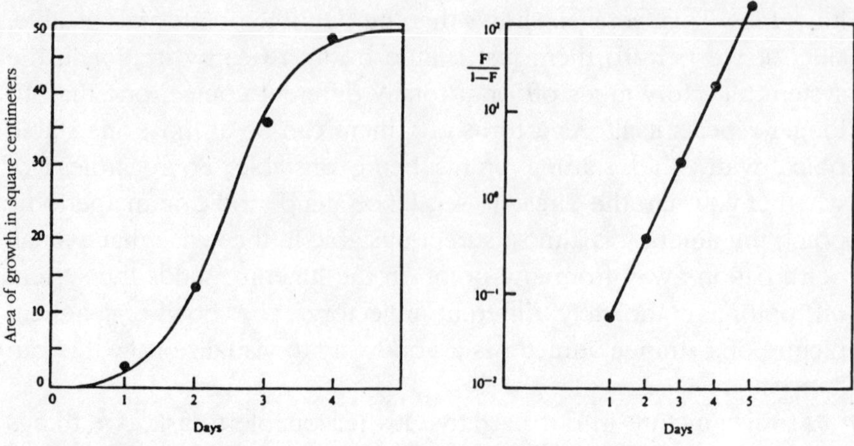

FIGURE 1.18. LOGISTIC GROWTH OF A BACTERIAL COLONY

and when this quantity equals 10^2, or 100, the system is at essentially 100 percent of its final size. This kind of plot is convenient, since any process whose behavior obeys the S-shaped logistic rule will appear as a straight line when plotted using the ratio $F/(1 - F)$.

Of course, logistic growth is just what we'd expect from a colony of bacteria with a fixed amount of nutrient, given the fact that such a colony can grow only as long as nutrient is available. We would expect a period of rapid growth at the beginning when there's more than enough nutrient to go around. When the nutrient starts running short, unrestricted growth is curtailed and the growth rate becomes more or less proportional to the current number of bacteria due to competition for the limited amount of nutrient. Finally, the growth tails off as the nutrient is exhausted.

What's interesting about Marchetti's work is the fact that this very same pattern seems to appear in areas where there is no obvious quantity available to play the role of the nutrient, or scarce resource. Figures 1.19 and 1.20 illustrate this point for the growth of mainframe computers in Japan and the cumulative production of Mozart's thirty-five major musical works. This latter illustration is particularly intriguing since it suggests that each of us has some kind of internal "program" regulating our creative output until death. Moreover, as the saturation point is approached, people seem to die when they have exhausted 90 to 95 percent of their potential, as measured by the limiting values of their productive curve.

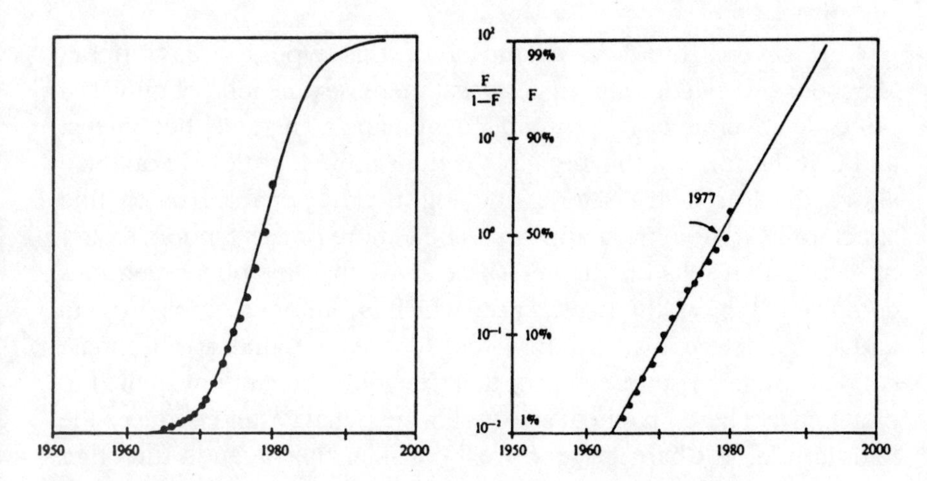

FIGURE 1.19. GROWTH OF MAINFRAME COMPUTERS IN JAPAN, 1950–2000

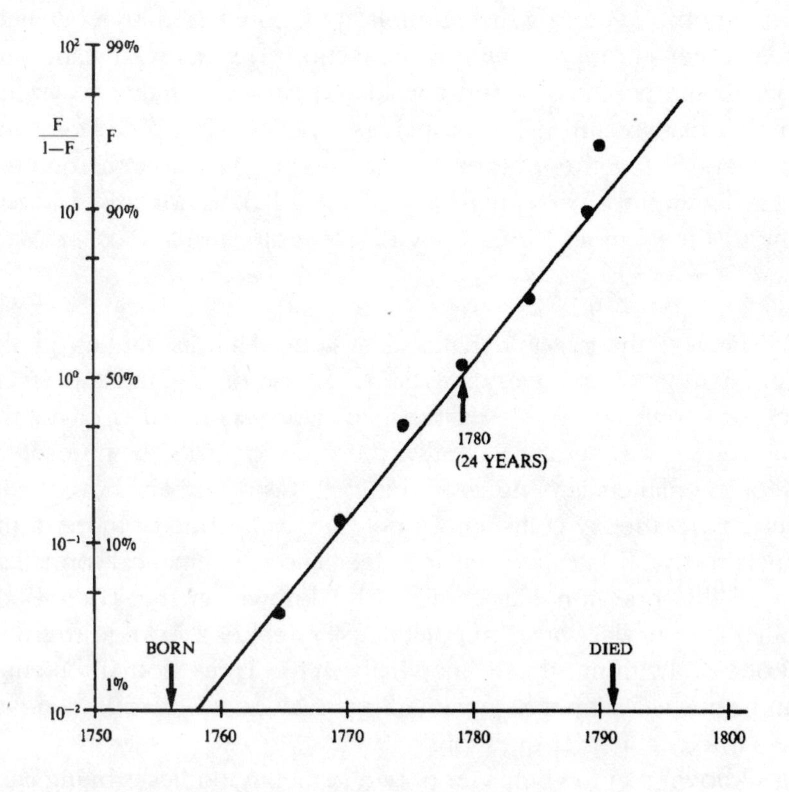

FIGURE 1.20. MOZART'S CUMULATIVE PRODUCTION OF MUSICAL WORKS, 1756–1791

With several hundred examples of this type to draw upon, Marchetti's work strongly suggests that there is some kind of universal principle governing a large number of human and natural phenomena, and that the form of this principle is the simple logistic rule outlined above. In Marchetti's setting, the logistic rule enables us to find structure in what initially appears to be a more or less random scatter of data. The results of Chapter Three show that this phenomenon is completely typical: the logistic rule, which is almost as simple in form and just as deterministic as the Circle-10 rule, can, under appropriate conditions, give rise to behavior that's every bit as random as the flip of a coin. So here Marchetti starts with the behavior and deduces the logistic rule; in Chapter Three we'll consider the situation the other way around, starting with the logistic rule and looking at the seemingly random behavior.

With these ideas of universal self-organization in mind, let's look at still another way in which complicated, counterintuitive behavior can emerge—through nonlinear interactions. Just as with static situations of the paradoxical sort considered earlier, dynamical systems can also display surprising properties when we couple elementary subprocesses together to form a larger system. Probably the quintessential example of how hard-to-understand behavior arises in this fashion is the famous Three-Body Problem of celestial mechanics.

The classical theory of dynamical systems had its origins in the determination of planetary orbits. And one of the most pressing questions confronting celestial mechanicians was whether or not the solar system was stable. Basically, the question asks, If the orbit of one of the planets is perturbed a bit, will this disturbance eventually result in a planetary collision? Or possibly cause one or more of the planets to fly off into interstellar space? The mathematical formalization of this question has come to be known as the Three-Body Problem, a puzzle whose complete answer is not known to this day. But one of the things that is known about the Three-Body Problem is that the answer cannot be obtained by breaking up the question into a sequence of simpler questions.

It's known that the behavior of two planetary bodies orbiting each other can be written down completely in closed form. Nevertheless, it turns out to be impossible to combine the solutions of three two-body problems to determine whether a three-body system is stable. Thus, the essence of the Three-Body Problem resides somehow in the linkages between all three bodies. And any approach to the problem that severs even one of these linkages destroys the very nature of the problem. So here is a case in which complicated behavior arises as a result of the interactions between relatively simple subsystems. And this is not a phenomenon confined to the nether world of planets and stars either, as the following example shows.

Conventional wisdom in the world of economics (Adam Smith's world, that is) says that in a system of goods and demand for those goods, prices will always tend toward a level at which supply equals demand. In short, the negative feedback from the supply/demand relationship to prices leads to a stable equilibrium point. Recently, maverick economists like Brian Arthur of the Santa Fe Institute have argued that this is not at all the way the real economy works. Rather, they claim that what we see is *positive* feedback in which the price equilibria are unstable. A familiar example will illustrate the point.

When video cassette recorders (VCRs) started becoming a household item a decade or so ago, the market began with two competing formats—VHS and Beta—selling at about the same price. By increasing its market share, each of these formats could obtain increasing returns since, for example, large numbers of VHS recorders would encourage video stores to stock more prerecorded tapes in VHS format. This in turn would enhance the value of owning a VHS machine, leading more people to buy machines of that format. So by this mechanism a small gain in market share could greatly amplify the competitive position of VHS recorders, thus helping that format to further increase its share of the market. This is the characterizing feature of positive feedback—small changes are amplified instead of dying out.

The feature of the VCR market that led to the situation described above is that it was initially unstable. Both VHS and Beta systems were introduced at about the same time and began with approximately equal market shares. The fluctuations of those shares in the early going were due principally to things like "luck" and corporate maneuvering. In a positive-feedback environment, these seemingly chance factors eventually tilted the market toward the VHS format until it acquired enough of an advantage to take over essentially the entire market. But it would have been impossible to predict at the outset which of the two systems would ultimately win out. The two systems represented a pair of unstable equilibrium points in competition, so that unpredictable chance factors ended up shifting the balance in favor of VHS. In fact, if the common claim that the Beta format was technically superior holds any water, then the market's choice did not even reflect the best outcome from an economic point of view.

This chapter has been a whirlwind tour through the land of observations, models, linguistic descriptions, dynamical systems, logical and visual paradoxes, unstable systems, randomness, algorithmic complexity, irreducibility and a whole lot more. So before moving on to a more leisurely consideration of science and surprise, let's try to summarize how these themes fit together to generate the unanticipated.

The economic example involving the dynamics of choice in VCR tape formats shows how paradoxical, unpredictable and just plain surprising behavior can emerge even in simple systems when the components of the system interact in ways that we don't fully understand. Sometimes the complex behavior is due to nonlinearities in which the outcome is disproportional to the input; sometimes, as in the VCR case, the problem lies with inherent, hidden instabilities in the system. In other cases the source of the counterintuitive behavior might reside in the dynamical rule itself, as seen in the chaotic behavior of the Circle-10 system. Paradoxes of various sorts also give rise to their own characteristic types of surprises, mostly on account of the presuppositions built into our linguistic descriptions of the world. These paradoxes are generally static versions of the type of counterintuitive behavior that comes out of unstable dynamical processes like the Circle-10 system. There are even situations, which we shall discuss in Chapter Four, in which surprises can arise when we ask if there is *any* computable rule governing a system's observed behavior. Taken together, these surprise-generating mechanisms—instability, deterministic randomness, uncomputability, irreducibility—are the strands from which we can hope to weave a science of surprise.

TWO

THE CATASTROPHIC

Intuition: Small, gradual changes in causes give rise to small, gradual changes in effects.

Look abroad through Nature's range
Nature's mighty law is change.

—ROBERT BURNS

In science, each new point of view calls forth a revolution in nomenclature.

—FRIEDRICH ENGELS

Mathematicians are a species of Frenchmen; if you say something to them, they translate it into their own language and presto! it is something entirely different.

—JOHANN VON GOETHE

CONTINUITY AND COMMON SENSE

In an editorial of April 2, 1985 addressing extinction events like the disappearance of the dinosaurs, *The New York Times* stated

> Terrestrial events, like volcanic activity or change in climate or sea level, are the most immediate possible cause of mass extinctions. Astronomers should leave to astrologers the task of seeking the causes of earthly events in the stars.

In this remarkably misguided view of the processes of both nature and science, the newspaper's editorial writers were arguing the implausibility of a dramatic, unexpected event like a meteorite impact being the proximate cause of the event that wiped out the dinosaurs sixty-five million years ago. With their implied predilection for smooth, gradual changes in earthly affairs, I wonder what these editors thought of the Black Monday stock market crash in October of 1987 or the Iraqi invasion of Kuwait in the summer of 1990! And, in fact, recent geological evidence argues against an earthly cause of the dinosaurs' demise, strongly suggesting instead that the long-sought "smoking gun" of that ancient meteorite strike is a large-impact crater at Chicxulub on the Yucatan peninsula.

But let's not be too hard on *The New York Times* editorial staff. Everyday common sense certainly does rest to a large degree on the notion of continuity and gradual changes. The basic idea is that the patterns, processes and structures of daily life don't change very much if they are distorted or disturbed by a small amount. So, for example, a minor delay in the subway train's arrival doesn't usually result in your being more than a few minutes late to work. And cranking up the temperature in the oven a few degrees higher than called for by the recipe does no great harm to the resulting carrot cake. In short, intuition and common sense say that small, gradual changes in causes give rise to small, gradual changes in effects. This fundamental principle underlies what's technically termed *structural stability*, a crucial property built in to most of the mathematical descriptions of natural phenomena we've inherited from classical physics.

To illustrate this basic idea, Newton's laws of universal gravitation tell us that the small disturbance to the Moon's position and velocity caused by a stray meteorite strike will eventually fade away—thus leading to no great change in the shape of the Moon's orbit. Similarly, Maxwell's equations of electromagnetism say that the patterns of the electric and magnetic fields generated by a moving charged particle do not change much if we make small changes in either the path or the charge of the particle. So, to a great extent, classical physics is the physics of structurally stable systems.

And a good thing, too, as it's hard to imagine how life could emerge and survive in a structurally unstable world, where every situation would differ dramatically from every other, and no pattern could be counted upon to repeat itself in a more or less regular manner. This kind of world would be totally chaotic, and hardly the sort of environment in which any sort of evolutionary process could get a foothold. So a certain degree of stability seems necessary for the very existence of life itself.

But not all phenomena of earthly concern are structurally stable. Nor are they continuous. For instance, the unpredictable outcome of a roll of the dice at the craps table or the turn of a card at the poker parlor both arise from processes in which a small change in the input (the position and velocity of the dice or the shuffling pattern of the cards) can lead to a big change in the final result. And most events in sports, like a goal in ice hockey or a completed pass in football, are likewise the end result of such a discontinuous process.

It's not just in the casino or on the playing field where discontinuity intrudes into everyday affairs. The same kind of processes also seem to lie at the heart of phenomena like price fluctuations on Wall Street, the breaking of waves on the beach and the outbreak of an infectious disease. In these situations, a seemingly minor change in the original circumstances—an investor's decision to buy a stock or an infected child contacting a playmate—can set off a major shift in the observed output, be it a market crash or a measles epidemic. Catastrophe theory is an attempt to go beyond the confines of classical physics by providing a mathematical framework for describing these types of discontinuous processes.

During the course of this chapter, the reader will necessarily be bombarded with a lot more concepts and terminology than those unfamiliar with mathematical phraseology and arcana are accustomed to digesting. So to provide a framework within which to assimilate this snowstorm of ideas from the world of dynamical systems and geometry, a brief, informal overview of what catastrophe theory is all about may provide a useful crutch for the reader to lean on.

Consider a system like the national economy. Suppose we're monitoring some measure of the performance of the economy, say the gross national product (GNP). This observed output of the economic system is determined by many factors—interest rates, employment levels, productive capacity and the like. We can think of the economy as a kind of machine; we feed in the value of each of these input quantities and the machine then produces a level of GNP as its output. Since the economy is a dynamical process, it's reasonable to consider the level of GNP as being a fixed-point attractor of the economic process. So for every value of the inputs, the economy moves to a particular level of GNP, which can be envisioned as a point in the space of states of the economy. And since every setting of the inputs produces such a point, there is a whole surface of GNP points that the economy may produce—at least one for every level of interest rates, money supply, production facilities and all the rest. Catastrophe theory is designed to study the geometrical structure of this surface.

Generally speaking, if we change the inputs just a bit, the corresponding level of GNP will also shift only slightly. But occasionally we'll encounter a combination of input values such that if we change them only a small amount, the corresponding output will shift discontinuously to an entirely new region of the GNP surface. Such a value of the inputs is called a *catastrophe point*. In colloquial terms, we might think of the catastrophe points as the straws that break the economy's back. As it turns out, these catastrophe points arise at just those input levels where there is more than one possible fixed point to which the system can be attracted. And the jump discontinuity is a reflection of the system's "deciding" to move from the region of one attractor to that of another. Catastrophe theory shows us that there are only a small number of inequivalent ways in which these jumps can take place, and it provides a standard picture for each of the different geometries that the surface of attractors can display. The reader should try to keep this simple picture of the goals of catastrophe theory in mind as we wend our way through the abstractions and applications that follow.

Catastrophe theory was announced to the general scientific community in 1972 with the publication of René Thom's remarkable book *Structural Stability and Morphogenesis*. The initial reviews in the most respected scholarly journals were extremely positive, containing statements like "Both Newton's *Principia* and Thom's book lay out a new conceptual framework for the understanding of nature" (Clive Kilmister in *The London Times Higher Educational Supplement*) and "it [the book] gives me a sense of liberation and enlightenment akin to what I imagine astronomers must have felt when offered Copernican heliocentric geometry . . . the sustained inspiration and the vast scope of the book put it firmly into the best tradition of natural philosophy" (Brian Goodwin in *Nature*). But less than five years later we find statements of the following sort appearing in equally prominent forums: "Exaggerated, not wholly honest . . . the height of scientific irresponsibility" (Marc Kac in *Science*) and "Catastrophe theory actually provides no new information about anything. And . . . it can lead to dangerously wrong conclusions" (James Croll in *New Scientist*). What gives here? How can equally eminent scientists come to such dramatically different opinions about a mere mathematical theory? What is it about catastrophe theory that led to statements of such lavish praise and heated outrage?

We'll answer this puzzling query at the end of the chapter, after we've had ample opportunity to see catastrophe theory in principle and in practice. But before jumping into the theory with both feet, a few words about models and reality are in order.

We laid great emphasis in the opening chapter upon the fact that the scientific answer to a question takes the form of a set of rules, rules having specific properties and modes of generation. Usually these rules are encoded in the form of a mathematical model or, more generally these days, a computer program. Moreover, the rules are used in two quite distinct ways: (1) to *predict* the outcome of future observations and/or (2) to *explain* past observations. Newton's laws of celestial motion are the quintessential example of a set of rules used for the first task, while the principle of natural selection in evolutionary biology exemplifies the second. Of course, in some cases the same set of rules can be used for both purposes, as with Newton's laws of motion. But this is generally not the case.

The central ingredient in the catastrophe controversy rests on a fundamental misperception of the nature of the rules the theory provides. Many practitioners of the mathematical arts seem unable to resist the temptation to use the mathematical pictures offered by catastrophe theory as a tool for prediction. Yet in most cases the mathematical structure and logic of the theory itself are intrinsically restricted to offering only explanations. As we shall see shortly, it's only in very special situations—mostly in physics—where catastrophe-theoretic ideas can be used to actually predict something in a quantitative way. For the most part, all we can hope to do is obtain the kind of qualitative guidance offered by Darwin's principle of natural selection. In other words, we can usually only get a picture of a situation, something more akin to an impressionistic painting than to a photograph. Much of the controversy surrounding the uses and abuses of catastrophe theory—especially in the social and behavioral areas—takes the form of rather hysterical outbursts against perceived attempts by practitioners to use the theory as a vehicle for prediction in situations where at best it can be used only to explain. Let's now illustrate these rather general ideas by looking at an example from recent world events showing catastrophe theory in action.

THE FALL OF THE WALL AND THE COLLAPSE OF A BEAM

On November 9, 1989, the Berlin Wall came tumbling down, symbolically marking the end of Communism in Eastern Europe. While viewed with equal measures of both happiness and relief by most Westerners, this event came about with an unplanned abruptness the world is still trying to understand and assimilate, especially in what has now become a reunified Germany. Interestingly, in 1979, exactly ten years before the fall of the Berlin Wall, Christopher Zeeman published a catastrophe-theoretic model for the shift of political ideologies that offers a mathematical glimpse at how such a discontinuous shift in political ideologies might actually come about.

Suppose we take as the basic aims underlying a society the ideals of the French Revolution: *liberty, equality, fraternity*. Moreover, since many argue that one of the main advantages of technological progress is that it will provide everyone with the chance for self-fulfillment, let's add a fourth ideal, *opportunity*. To some degree, these four aims fall into two conflicting pairs of social goals: an economic conflict between equality and opportunity, a political conflict between liberty and fraternity.

Clearly, different individuals in a society will place different emphasis on these underlying goals. So let's assume that each individual's emphasis can be measured using two independent quantities, a and b. For operational convenience, let's assume these variables are scaled to lie between 0 and 1. The economic parameter a measures the relative emphasis the individual places on opportunity versus equality, while the political parameter b measures the relative emphasis placed on fraternity as opposed to liberty. So we are assuming that the ideological position of each individual can be represented by a point in a two-dimensional space of parameters P, whose elements consist of pairs of numbers (a, b). This space of parameters is shown in Figure 2.1.

Now we want to plot the opinions of individuals on societal matters such as censorship, free speech, unions, socialized medical care and free education. Suppose there are n such issues phrased as questions and that the answers given by any individual are expressed on a scale ranging from "strongly prefer" to "strongly against." Further, assume that we can associate these responses to some numerical measure of preference. The n questions then determine the axes of an n-dimensional space X, so that the answers given by each individual can be represented by a point x in this space.

With the above conventions, each person in the society can be represented both as a point in the parameter space P, representing that individual's views on the overall economic and political goals of the society, and as a point x in the space of opinions X, representing the individual's views on specific social issues. Since P is two-dimensional and X is n-dimensional, each person's views can be represented by a single point in a space of dimension $n + 2$. Geometrically, the society as a whole can then be pictured as some great cloud of points in this space. We'd like to know more about the shape of this cloud.

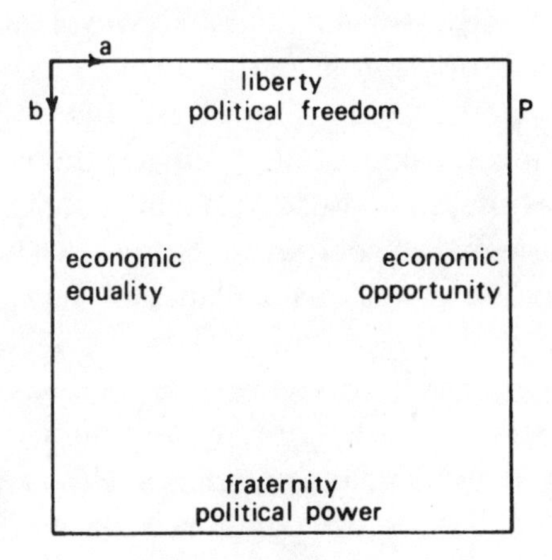

FIGURE 2.1.
THE SPACE OF INDIVIDUAL IDEOLOGIES

Let's assume that for each point p in the parameter space P (i.e., for each individual who holds to a particular set of ideals), there is a distribution of opinions on the social issues such that these opinion points in the space X tend to cluster near a point x_p. In other words, if individuals share a similar emphasis upon the four basic aims, their opinions on social matters tend to be close together. We can think of these points x_p as being the places where the probability of a certain response to the questions is higher than at any nearby point. But note that there may be several such local maxima.

Sociologically speaking, we can interpret each such local maximum point as a particular kind of stable "ideology," in the sense that the opinion represented by such a point x_p is shared by a large segment of society. We call it stable because a small change in a person's ideals from p to p^* doesn't change that individual's response to the questions very much; hence, the response point x_{p^*} is very near the original point x_p. Most of the major political parties in the Western world exemplify stable ideologies in this sense. On the other hand, points where the opinion profile is a local minimum correspond to unstable ideologies for obvious reasons. A political party like, say, the short-lived Symbionese Liberation Army of Patty Hearst fame is a good example of an unstable ideology, as it wouldn't take much by way of change in its members' opinion profiles to dramatically change their responses to the social questionnaire.

Mathematical arguments show that the set of all such critical points x_p (ideologies) forms a two-dimensional surface, which for notational convenience we shall denote here as S. Clearly, as we shift the social goals (ideals) away from the point p a little bit, the corresponding ideology also usually moves only a small amount. However, at some values of p there may be an abrupt shift from the current ideology to another. This happens when the change in goals causes a local maximum point to cease being a maximum. Catastrophe theory gives us a tool by which to study how these ideological shifts come about as a consequence of variations in the societal goals p.

Under the foregoing conditions, the mathematics of catastrophe theory assures us that the geometry of the surface S of ideologies almost always looks like that shown in Figure 2.2. We'll say more about this "almost always" part a bit later. For now, let it suffice to say that this is the typical, or *generic*, geometry governing such a situation. Note the shaded regions N_1 and N_2 in the parameter space of social

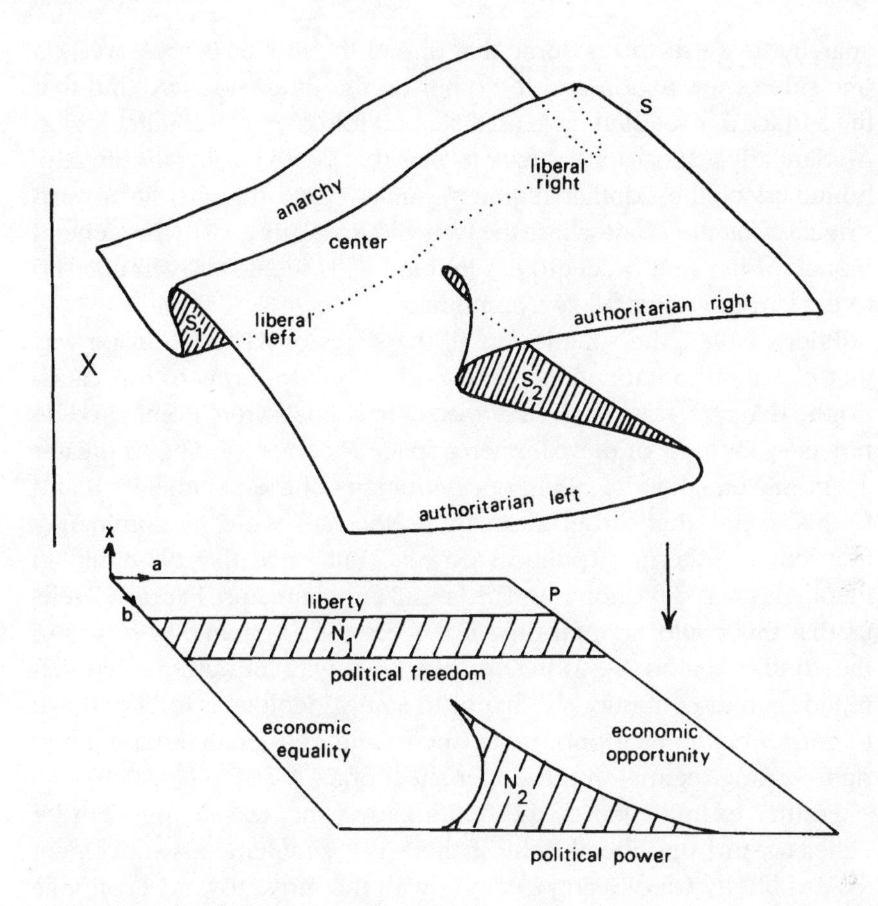

FIGURE 2.2. SURFACE OF IDEOLOGIES

ideals. These are what we might call *conflict regions*, in the following sense. Suppose that p is the profile of an individual on the boundary of a conflict region. At this point the probability distribution of answers to the questions about societal issues has two stable local maxima (i.e., two stable ideologies), represented by points on the upper and lower sheets of the surface S. These maxima are separated by an unstable local minima in the shaded region of S, which corresponds to a centrist-oriented authoritarian regime—a political ideology not likely to persist in the world of *Realpolitik*. Thus, the surface is two-sheeted over a neighborhood of any such conflict point p.

As we leave either of the conflict regions, one of the two modes is preserved and the other has to disappear. For example, the current ideology in Russia seems to be a transition from the region N_1 to anarchy. If we further assume that one of the modes is preserved on one side of the region and the other on the other side, we find that the surface S must then have an S-shaped fold over the conflict region N_1. Similarly, catastrophe theory tells us that the two folds forming the boundary of the conflict region N_2 must come together in a cusp singularity at the point where the two fold lines meet. With this general model for the shift of ideologies in mind, let's see how it can be used to explain the downfall of Communism.

Prior to 1989, the situation in all the states of Eastern Europe was that of an authoritarian-left political ideology. In terms of our catastrophe diagram, this means that the societal goals were dominated by points in the part of the parameter space P corresponding to greater economic equality. Recalling our definitions of the parameters a and b, this means that the goals of these societies were predominantly focused on fraternity (political power). But since the 1989 shift in ideologies was discontinuous (i.e., rapid and dramatic), Figure 2.2 tells us that this could happen only if the economic parameter a was in the conflict region N_2. Moreover, the only way an authoritarian-left regime can discontinuously change to a new ideology is for it to move to greater economic opportunity (increasing a), thereby jumping to a right-leaning regime on the upper sheet of S.

Further examination of the figure shows that this change can be either toward the liberal-right, if there is a simultaneous movement toward liberty (decreasing b) along with the move toward economic opportunity, as in modern-day Hungary or Poland, or toward an authoritarian-right regime like Singapore, if the political mood is focused primarily upon opening up economic opportunities. But the one thing that cannot happen is a rapid, discontinuous shift from an authoritarian to a liberal regime, because the only such transition is a smooth path on S in the direction of decreasing economic equality (i.e., decreasing b). And, of course, this impossible jump is exactly what we've seen national leaders trying in vain to bring about over the past few years in, for example, many of the former Iron Curtain countries.

This political ideology example underscores all of the main assumptions, strengths and weaknesses of catastrophe theory. It's worthwhile to pause here for a moment to summarize these various facets of the theory. But since catastrophe theory is a mathematical theory, to do this we shall have to temporarily forsake the real world for the world of abstractions. We'll get back to terra firma as soon as these general catastrophe-theoretic ideas are firmly in hand.

In the opening chapter, we saw that there are three basic types of attractors for dynamical systems: fixed points, periodic orbits and strange attractors. What's called *elementary* catastrophe theory (the only part that's really in good mathematical shape) deals directly with only those systems whose attractors are fixed points. For such systems, each value of the input parameters determines at least one fixed-point attractor to which the system will try to move. So as we run through all possible parameter values, a *surface* is generated, each of whose points is an attractor of the dynamical process. As stated earlier, we can think of catastrophe theory as a way of classifying the possible geometries of this surface of fixed points. What the theory's founder, René Thom, called a *catastrophe* (apparently the French *catastrophe* is not quite so catastrophic as the English *catastrophe*) corresponds to those parameter values where the fixed point governing the system's behavior shifts from being a stable attractor to an unstable one. This is how a small change in something (a parameter value) can lead to a discontinuous shift in something else (the particular fixed point to which the system is attracted).

Of crucial importance in this setup is the notion of a *family* of functions. Let's illustrate the idea in a particularly simple setting. Suppose we consider the set of functions whose graphs are shown in Figure 2.3. Observe how curve (b) passes through the origin horizontally. However, an arbitrarily small perturbation of this curve near the origin can lead to the nearby curve (c), which has two humps (a local minimum and a local maximum), or to curve (a), which has no humps at all. Note, though, that a similarly small distortion of either curve (a) or curve (c) near the origin preserves the geometric character of those curves—two humps, or critical points, for curve (c) and none for curve (a)—rather than destroying the hump structure as was the case for curve (b). Thus, what we have here is a family of curves that is stable *as a family* since the geometric character of the critical point at the origin remains unchanged when almost every member is jiggled just a little bit. But there is one "black sheep" member of the family, curve (b), which is unstable when considered as an individual since its geometric structure does change if it's disturbed even a little bit.

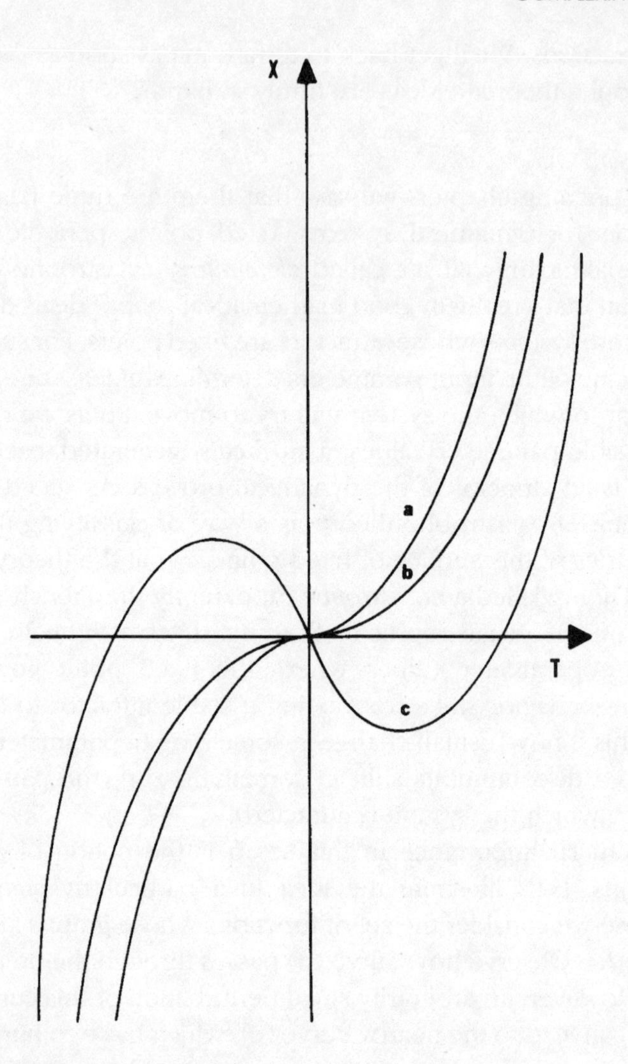

FIGURE 2.3. A FAMILY OF FUNCTIONS

To see this stability issue more clearly, think of curve (b) as being an elastic band. If you push just a little on the band at the origin, it may distort to either curve (a), having no critical points (humps) or to curve (c), having two such points. It all depends on exactly how you push on the band. But if the band starts in the shape of curve (a) or curve (c), a small enough push cannot create or destroy the curve's geometric structure. This is why we call curve (b) unstable and curves (a) and (c) stable.

In our political ideology example, the relevant family of functions is the set of probability functions characterizing the distribution of answers to the social questions. These functions are labeled, or named (technically, *parameterized*), by the points p of the social goal space. So for every profile p of social goals or ideals, there is a particular probability distribution describing the likelihood of the answers given by people having that profile p. The unstable members of this family correspond to those distributions for which the opinion point x_p ceases to be a local maximum when we vary p, thereby forcing the dominant ideology to shift discontinuously to a new point on the ideological surface.

So to summarize, catastrophe theory deals with the fixed points of families of functions. The catastrophes occur when, as we move in a continuous way through the family—usually by smoothly changing parameters describing the system—a stable fixed point of the family loses its stability. This change of stability forces the system to move abruptly to the region of a new stable fixed point. Note that in the ideology example it was not necessary to know the exact form of the probability functions forming the family. We needed only to assume their existence, at which point catastrophe theory tells us the possible types of geometries that the surface of fixed points can have. Nevertheless, it seems reasonable to assume that if we know the precise mathematical form of the family members, we can say a lot more about the situation, perhaps even being able to make some quantitative predictions about what the system will do. Sometimes we do have such knowledge, especially when the system under investigation is one from physics. Let's now return to the real world and look at just such a situation.

Buckling Beams

In the opening chapter we saw a tragic example of how mechanical structures can fail when we considered the seemingly paradoxical collapse of a Kansas City hotel balcony. Here we consider a very simplified version of this situation, involving the buckling of a column or beam.

Suppose we have an elastic strut of length l subject to a force K exerted at each of its ends. The inherent symmetry in this situation will be destroyed due to manufacturing imperfections, causing the beam to buckle either upward or downward as K is increased. If the beam buckles upward and we agree to measure the amount of buckling by the quantity x, we end up with the situation shown in Figure 2.4.

Assume now that there is a load L applied to the center of the strut. Then the displacement x will decrease continuously until the load reaches a critical value at which point the strut will suddenly jump from being buckled upward to a downward-buckled state. Catastrophe theory enables us to study these discontinuous shifts as we smoothly vary the two parameters K and L. Figure 2.5 shows the various possibilities.

Initially, the beam is in an unbuckled state ($x = 0$). As the two parameters change along the numbered path in Figure 2.5, the strut behaves in the following manner: nothing happens as K increases until the cusp point is reached, whereupon the strut begins to buckle upward. More than two hundred years ago, the Swiss mathematician Leonhard Euler showed that this occurs when $K = \pi^2\lambda/l^2$, the quantity λ serving to measure the elasticity of the strut. As K is further increased along path 2, the strut buckles more sharply. If we now keep K constant and increase L along path 3, the displacement gradually lessens until at the point P' it jumps onto the other sheet of the behavior surface, which corresponds to a sudden snap into the downward-buckled state. If the load L is decreased toward point 4, the amount of buckling x changes continuously, the strut maintaining its downward-buckled state past the point P'. In other words, it does not return to the upper surface until it reaches Q', at which point it suddenly snaps into the upward-buckled state.

You might well ask, How do we know that the geometry shown in Figure 2.5 is the right one for this situation? The answer ultimately rests with the precise mathematical form of the function characterizing

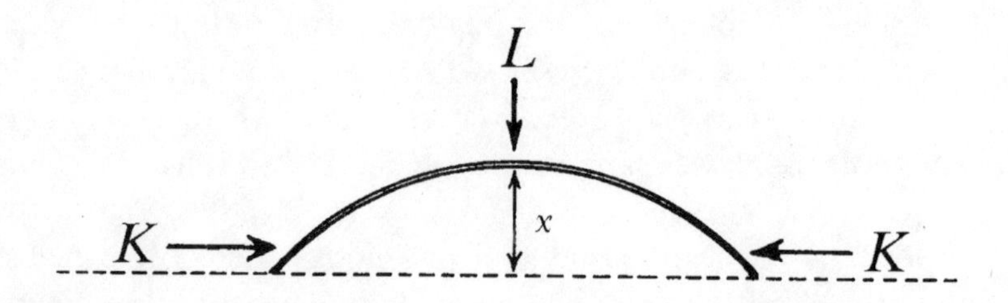

FIGURE 2.4. AN ELASTIC STRUT

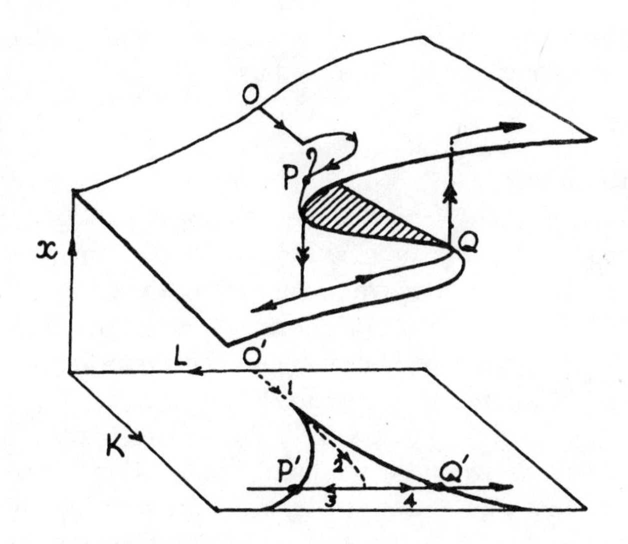

FIGURE 2.5. GEOMETRY OF BEAM BUCKLING

the total energy of the system. Readers interested in the details of this energy function are invited to consult the To Dig Deeper section for this chapter. Classical physics tells us that the beam will always seek a state that minimizes the total energy. Since the expression for the energy defines a *family* of functions, each member of the family being characterized by a particular value for the two parameters K and L, we have the same situation as for the political ideology example—with the notable difference that for the beam we actually know the mathematical form of each member of the family. The machinery of catastrophe theory then allows us to assert that the geometry shown in Figure 2.5 is indeed the one governing the behavior of the beam.

Actually, catastrophe theory tells us much more. It asserts that the geometry of Figure 2.5 is *universal* for all systems in which two input parameters govern the behavior of a single output variable. This startling fact is but one of the conclusions that can be drawn from the famous Classification Theorem of elementary catastrophe theory.

THE MAGNIFICENT SEVEN

Take a sheet of ordinary notepaper. Crumple it up any way you wish, subject to the sole proviso that you don't crease it. In other words, the twists and turns in the paper must all be what mathematicians and laymen alike would term *smooth*. Now pretend you're a fly walking around on the surface of this crumpled-up wad of paper. What would you see? The answer to this question sparked off one of the major developments in modern mathematics, the theory of singularities of smooth mappings.

In 1955, Hassler V. Whitney showed that when you scrunch up a piece of paper without putting any creases into it, any point on the resulting wad is one of three types: (1) a *regular* point, around which the surface is flat, (2) a *fold* point, where if you move away from it just a bit in the wrong direction, you fall off an edge, and (3) a *cusp* point, which is where two lines of fold points come together. The overall situation is shown below in Figure 2.6. It's important to note here that what's shown in the diagram is only the *local* geometry, a kind of magnified version of a small part of the surface of the paper. Globally, the wadded-up paper may consist of many such regions of regular, fold and cusp points.

In more formal terms, we can regard the above paper-crumpling exercise as simply a mapping that takes points of the original flat sheet of paper onto the points of the two-dimensional surface of the crumpled ball of paper. Whitney's result, that there are only three fundamentally different ways to wad up the paper without creasing it, serves as a role model for what René Thom was trying to do for surfaces of dimension greater than two.

In contrast to our earlier examples of ideological revolutions and bending beams, the paper-folding problem doesn't seem to involve any parameters; hence, there doesn't seem to be any natural *family* of functions involved here. So where does the *family* of functions needed for catastrophe theory come from in this paper-crumpling situation?

The answer is fairly straightforward, residing in the fact that there are many different ways to smoothly deform a sheet of paper. Each of these ways can be thought of as a different mapping transforming a plane (the original flat sheet of paper) into a wadded-up piece of paper. But this wad is just another plane, topologically speaking, since it can be smoothed out into a flat sheet. Hence, the family of functions is simply the collection of all such ways to smoothly crumple a piece of paper (i.e., ways to smoothly map a flat surface like a plane to a plane). The generic way results in a planar surface having only regular, fold and cusp points. These mappings are the stable members of the

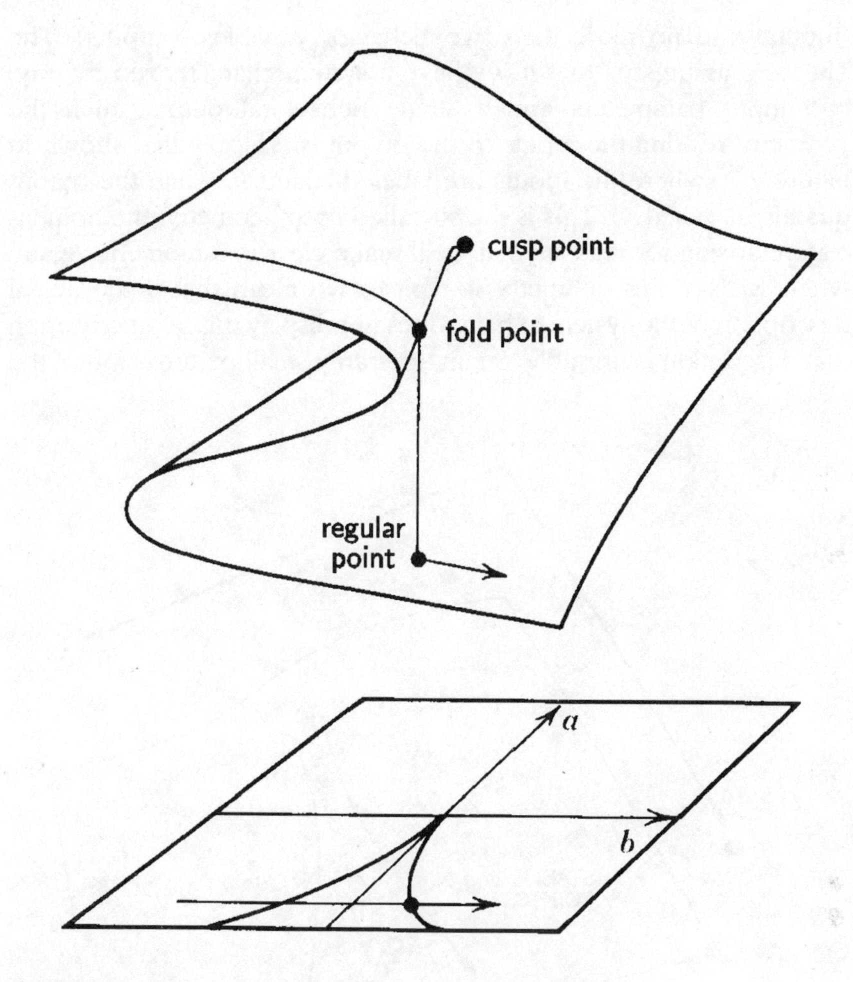

FIGURE 2.6. THE LOCAL STRUCTURE OF A CRUMPLED SHEET OF PAPER

family. But as we run through the different ways of twisting, folding and bending the paper, we will occasionally encounter *singular* transformations giving rise to more complicated types of points, like a pleat. However, the slightest change in these atypical transformations leads back to one of the "nice" crumplings. Thus, the singular transformations are unstable.

The Classification Theorem

The celebrated Classification Theorem tells us the generic ways to move through a function family characterized by up to six parameters (inputs) and no more than two behavior variables (outputs). The Theorem assures us that if we have a system characterized by, say, two input parameters and a single behavioral output, then the geometry relating the inputs to the output is typically that shown in Figure 2.7 (where the inputs are labeled *a* and *b*, while the output quantity is called *x*). This is the so-called cusp geometry, the nomenclature arising for reasons that we'll make clear in a moment. Again, when we say this geometry is typical, we mean that if the actual description of the system at hand does not display this geometry, then that description is unstable. So an arbitrarily small perturbation of the

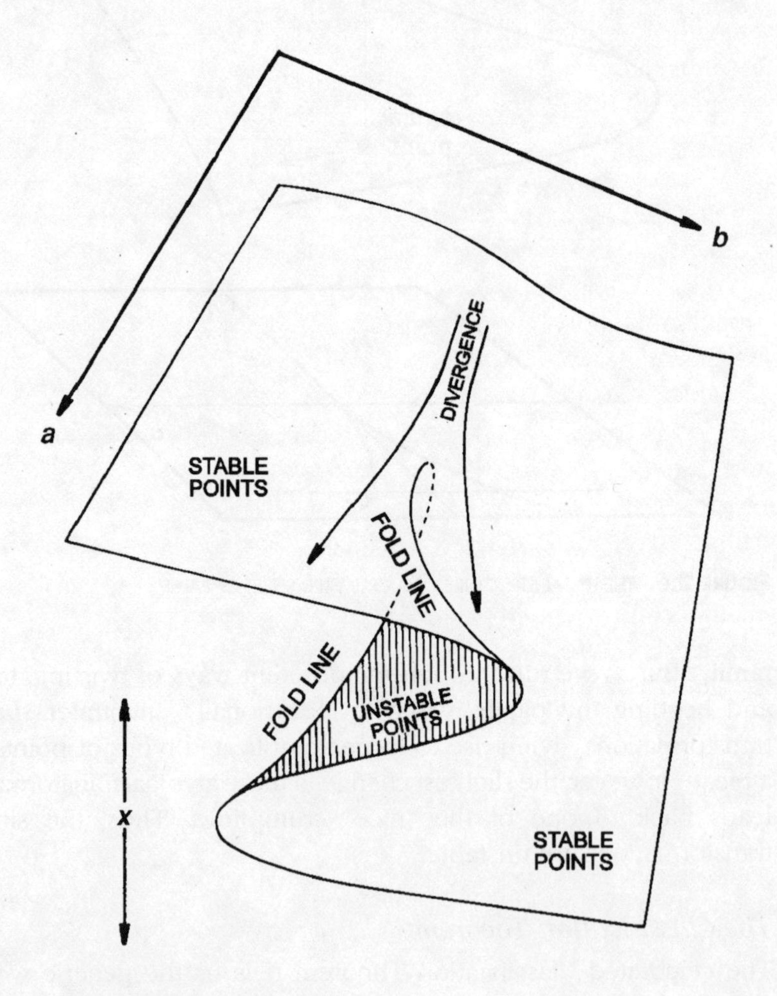

FIGURE 2.7. THE CUSP GEOMETRY

original system will take us to a stable description. And the cusp geometry of Figure 2.7 is a universal picture of all stable descriptions of systems having two inputs and one output.

To see how the Classification Theorem arrives at such far-ranging conclusions, consider again the function family shown earlier in Figure 2.3. This diagram is actually an abridged edition of a complete diagram that would show a family containing an infinite number of functions, one member for each value of a single parameter a. In more colloquial terms, we might think of each value of a as being the "name" of one of the functions in the family. So as we let a take on different values, we move through the members of this family, one by one, going from members having no critical points (humps) like curve (a) to family members like curve (c) having two critical points. As already noted, both these types of functions are stable in the sense that if we perturb them just a bit (i.e., change a by a small amount), we come to a nearby member of the family having the same number of humps. So in this entire family of functions, only curve (b) is atypical, or nongeneric, in this sense.

A lot of high-powered mathematics, mostly from the field of singularity theory, shows that *any* family of functions parameterized by a single variable can be transformed by a smooth change of variables to exactly the family of descriptions shown in Figure 2.3. So this is a universal, or *canonical*, picture of how one input can give rise to one output. The overall geometry for this situation, termed the *fold*, is shown in Figure 2.8. Note that the unstable member of this family corresponds to the member whose "name" is $a = 0$. The point $a = 0$ is called a catastrophe point. For positive values of a, the family members have no critical points, while for negative values they have two critical points—one a local minimum, the other a local maximum.

The Classification Theorem gives us the standard geometry for all functions having at most six input parameters and one or two output variables. The seven canonical geometries for up to four inputs are listed in Table 2.1. The table gives only the pet names for these various cases rather than the explicit mathematical formulas characterizing each standard geometry. Moreover, since Thom's original result only went up to four inputs, our listing stops there as well. For the complete, extended classification together with its associated mathematical finery, we invite the interested reader to consult the references cited in the To Dig Deeper section.

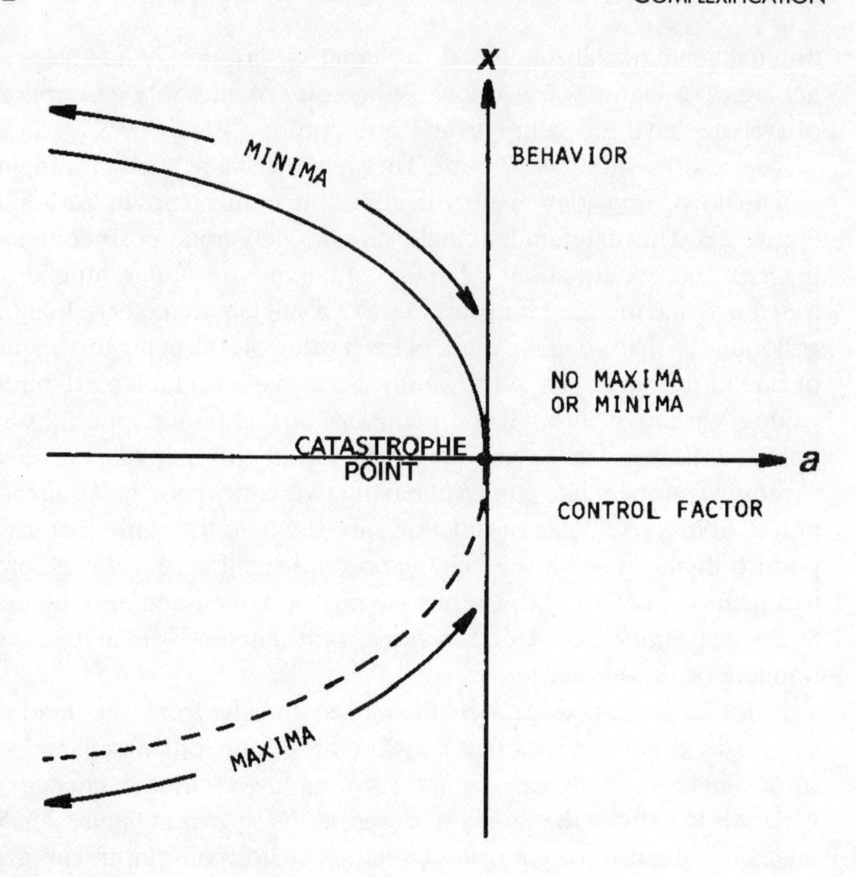

FIGURE 2.8. THE FOLD GEOMETRY

Examining the mathematical mandala of Table 2.1, the right questions for the reader to be asking about now are, What do all these mystical incantations about folds, cusps, butterflies, umbilics and all

TABLE 2.1.
SEVEN ELEMENTARY CATASTROPHES

Number of Inputs	*One Output*	*Two Outputs*
1	Fold	—
2	Cusp	—
3	Swallowtail	Hyperbolic umbilic
		Elliptic umbilic
4	Butterfly	Parabolic umbilic

the rest really *mean*? Where do these colorful terms come from? And how can they be used to answer interesting questions about real-world situations? Let's first settle the terminological issue.

Cusps and Butterflies

Suppose we take the cusp surface M shown in Figure 2.7, putting the plane of the two parameters a and b beneath it. Now project the surface M vertically down onto this plane. Under this projection, the two fold lines map into the two branches of the curve C in the plane, while their intersection, the cusp point, corresponds to the parameter values $a = b = 0$. The entire situation is shown schematically in Figure 2.9.

The reason this situation is called the cusp is that the mathematical expression for the curve C is a formula that algebraic geometers term a cuspoid, since it involves two curves coming together in a sharp, spikelike intersection that looks like what in everyday language is called a cusp. And the fact that it's necessary to "jump" or "fall" from

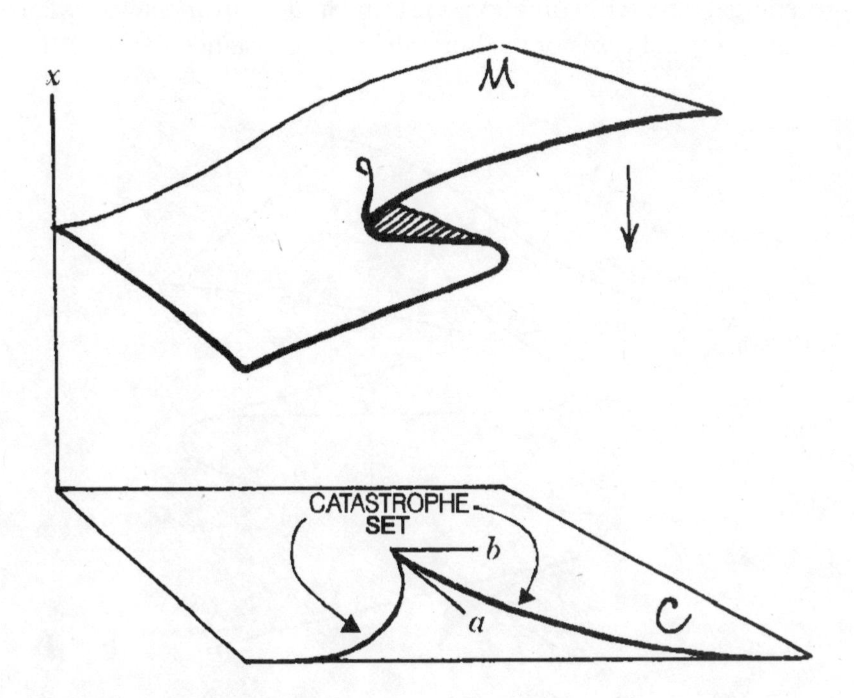

FIGURE 2.9. THE CUSP GEOMETRY

one sheet of the surface to the other when crossing a fold line gives rise to mathematical (although not necessarily real-world) "catastrophes." So the entire picture of Figure 2.9 is, by abuse of language, often termed the *cusp catastrophe*. Just to cement these terminological ideas in place, let's look at one of the higher-dimensional catastrophes, the butterfly.

As noted in Table 2.1, the butterfly catastrophe involves four input parameters. But since we have not yet mastered the delicate art of drawing in four dimensions, it's impossible to represent the complete situation pictorially in one fell swoop as for the cusp. So we'll have to content ourselves in this case with depicting only sections of the butterfly surface determined by subsets in the space of the four input parameters. Let's call these parameters a, b, c and d, and begin by considering the situation when a is negative and b is zero. The butterfly surface for this case, along with its projection onto the parameter plane determined by c and d, is shown in Figure 2.10. We see immediately from this projection why this situation is called the butterfly catastrophe.

The reader should note in the figure how all the catastrophes preceding the butterfly on Thom's list (the fold, cusp and swallowtail) appear as special cases of the butterfly. So, for instance, if we section

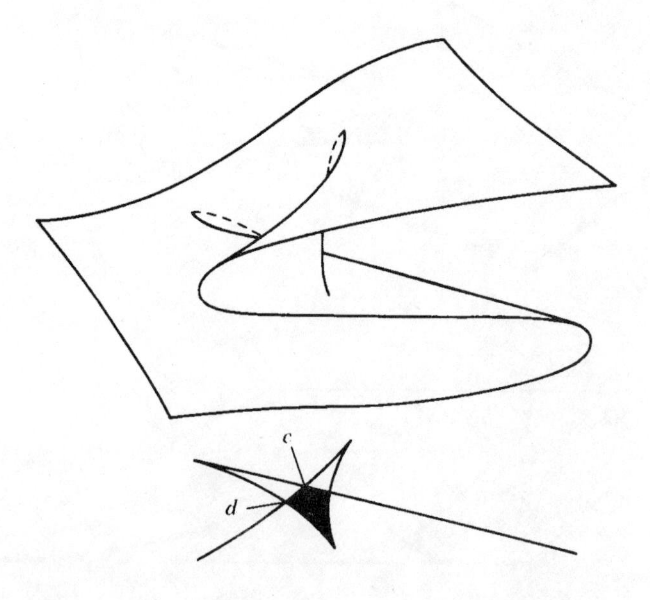

FIGURE 2.10. THE BUTTERFLY GEOMETRY FOR $A < 0$, $B = 0$

the butterfly surface by fixing the values of c and d equal to zero, the butterfly reduces to the cusp. In a similar way, sectioning the cusp manifold of Figure 2.7 by setting b equal to zero leads to the fold catastrophe. So as we add input parameters, the old behavioral modes are retained as special cases of the new patterns that appear.

As we saw in the political ideology example considered earlier, what's important about the butterfly is the possibility for a third mode of stable behavior to emerge, above and beyond the upper and lower sheets possible with the cusp. This occurs in the blackened *pocket-of-compromise* region shown in Figure 2.10. By varying the parameters c and d, this stable region of compromise behavior may be broadened, narrowed or made to disappear entirely. The various possibilities are indicated in Figure 2.11, which shows cross-sections of the complete butterfly catastrophe set.

While these matters are of great interest to mathematicians, it's hard not to wonder just a bit how this discussion of stable and unstable functions, catastrophes, cusps, butterflies and all the rest connects up with everyday phenomena like wars, water waves and collapsing beams. So with this mathematical armada at our beck and call, let's turn to the matter of linking catastrophes to life.

PHYSICS AND METAPHYSICS

Consider a bundle of light rays emanating from a point source. As these rays move, they encounter some medium—a mirror, a lens, an opaque object—that reflects, refracts or obstructs them. One of the most common ways this pencil of rays can be redirected is to be reflected off a mirror. Or light rays shining through a kitchen window might bounce off the inside of a coffee cup, as shown in Figure 2.12. The pattern of light reflected off the inside of the cup is what's termed a *light caustic*. By making use of Fermat's Principle, which asserts that light rays travel from one point to another along a path of minimal time, these caustics can be calculated and explained using catastrophe-theoretic arguments. Let's sketch the basic idea.

First we write down the algebraic expression for the length of the path from the light source to an arbitrary point on the surface of the coffee inside the cup, assuming the ray reaches this point after having been reflected from a point on the inner surface of the cup. Fermat's

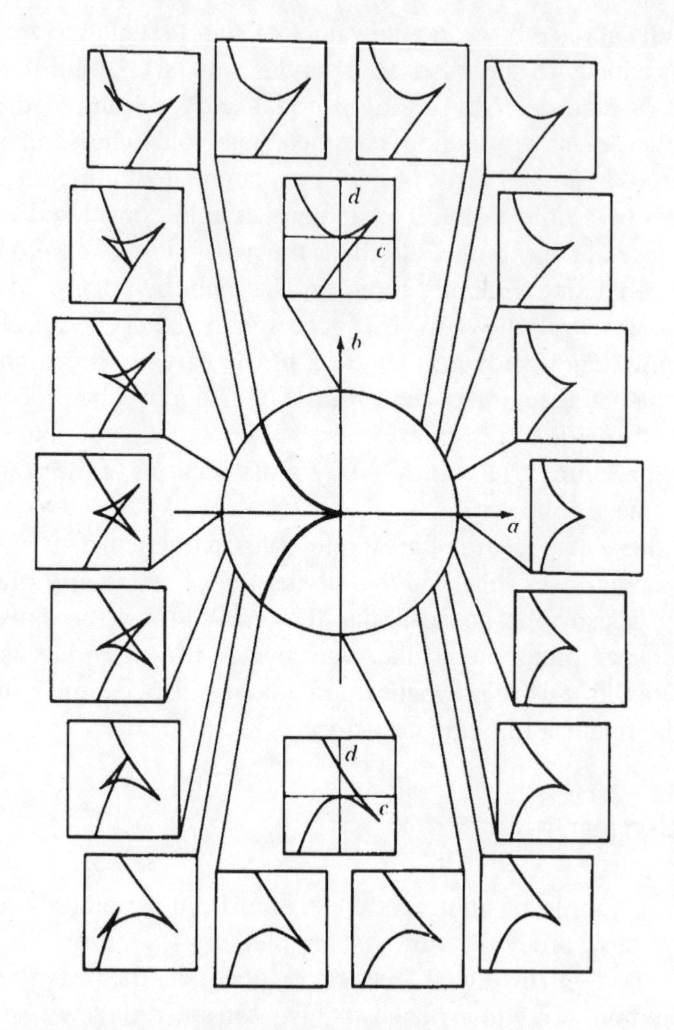

FIGURE 2.11. THE COMPLETE BUTTERFLY GEOMETRY

Principle then says that the path the ray travels from the source to the point on the coffee's surface via the reflection point on the cup must be as short as possible. Alternatively, we can invoke Snell's Law, stating that the angle the light ray makes when it strikes the inner surface of the cup must equal the angle it makes upon leaving the surface (i.e., the angle of incidence equals the angle of reflection). Both approaches lead to the picture shown in Figure 2.13. Employing a modicum of algebra, we find that the equation for the half-moon-

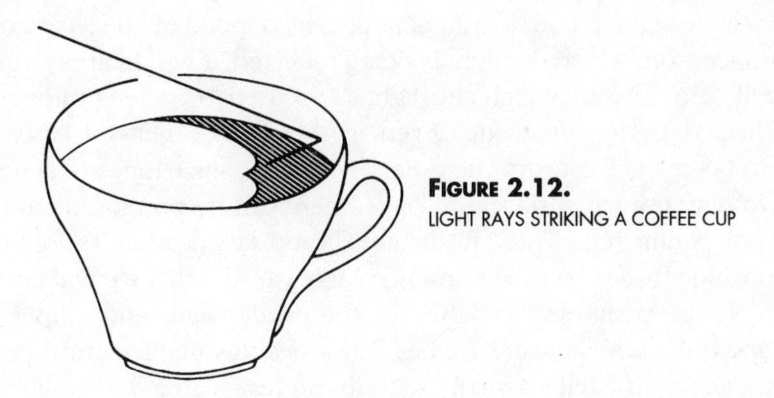

FIGURE 2.12.
LIGHT RAYS STRIKING A COFFEE CUP

shaped arc (i.e., the caustic) also describes the fold lines of the cusp catastrophe, shown earlier in Figure 2.9.

So here we're in much the same situation as in our earlier example of the bending beam. We have a function describing the behavior of the system, together with a principle of physics asserting that the states we will observe coincide with the places where that function takes on its extremal (i.e., largest or smallest) values. And since these functions involve a single behavioral variable (the point on the inner surface of the cup/the deflection of the strut) and two input parameters (the coordinates of the point on the coffee's surface/the force and load on the beam), we suspect that the governing catastrophe is the cusp. The mathematics of catastrophe theory allows us to validate this conjecture.

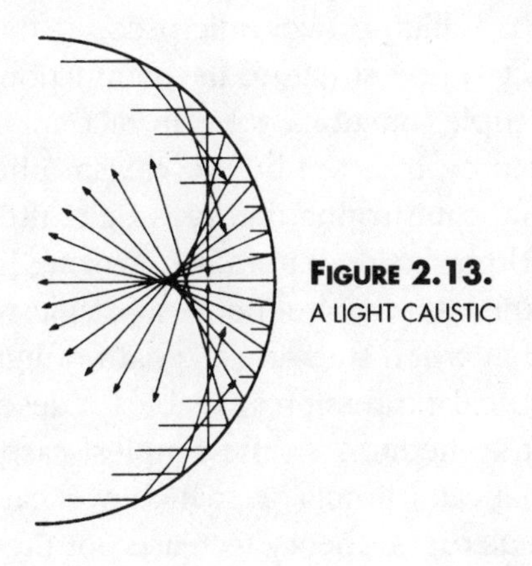

FIGURE 2.13.
A LIGHT CAUSTIC

The foregoing line of argument doesn't depend on the shape of the surface from which the light is being reflected; it would apply equally well to a case in which the light is focused by a lens rather than reflected by a coffee cup. Even in these more general cases, the caustics are still catastrophe points. However, since they are generally two-dimensional surfaces in space, there can be no more than three input parameters. Thus, from the Thom Classification Theorem we conclude that the only structurally stable caustics that we will ever see are folds, cusps and sections of the swallowtail, and elliptic and hyperbolic umbilic catastrophes. But does this mathematical prediction jibe with what's actually seen in the real world?

In laboratory experiments, Michael Berry of Bristol University in the U.K. has examined this question by looking at caustic junctions where several bright lines of light come together, like the rippling patterns seen on the bottom of a swimming pool. A casual glance suggests that these are three-way junctions, a pattern often found in nature in things like cracks in dry lake beds and the branching structure of the veins on many types of plant leaves. But the mathematics of catastrophe theory argues that this kind of pattern is *very* unusual, in the sense that it is structurally unstable; no catastrophe surface has a cross-section displaying such a pattern. So what's the story? Do these triple junctions occur regularly in nature? If so, the predictions of catastrophe theory must be wrong. Or can it be that catastrophe theory is right, and it's our observations that are wrong?

To settle this matter, Berry and his associate J. F. Nye set up equipment to produce and photograph caustic junctions in the lab. Looking through a high-power microscope, they discovered previously unsuspected fine structure in the junctions. What had earlier appeared to be triple junctions were in fact curved triangles having exactly the structure predicted by the catastrophe-theoretic analysis. So it turns out that catastrophe theory is right after all when it comes to describing and predicting naturally occurring light caustics.

What makes the catastrophe-theoretic magic work so well in this case is the fact that when studying the path of light rays, we have an explicit mathematical expression available for describing how the light moves through the medium. In the simplest case, Fermat's Principle gives a mathematical formula to which we can apply the formal operations of catastrophe theory to tease out the patterns and paths that the light rays will follow. This is an example of what's often termed the *physical way* of catastrophe theory: take an explicit mathematical expression for the phenomenon under study, and apply the methods of singularity and catastrophe theory to reduce this representation to the canonical geometry (or geometries) that the mathematics ensures must characterize the process.

The basic question here, of course, is Where did that mathematical representation come from in the first place? Once you have it, the tools and techniques of catastrophe theory can help in squeezing out every last bit of information from it. But how do you get it? And what, if anything, can be done if you don't know the governing mathematical relations beforehand? As a way station on the road to answering this query, let's move through the territory of the biologist and talk about how cells perform the miraculous task of organizing themselves into organisms.

Catastrophes and Morphogenesis

The greatest unsolved problem in biology, the real terra incognita of the field, is the enigma of embryology. How is it that an initially homogeneous ball of cells can differentiate and organize itself into one of the myriad species of living things we see roaming the four corners and swimming the seven seas of the Earth today? This is the problem of cellular differentiation and morphogenesis, or the emergence of form. Stimulated by work on morphogenesis in the late 1950s by the British developmental biologist C. H. Waddington, René Thom originally developed catastrophe theory as a mathematical way of addressing this very question. And to this day catastrophe theory remains the leading candidate to provide a rational explanation for how the local genetic coding in individual cells could cause the global unfolding of the embryo. As Thom once put it, "In developmental biology, how could they [biologists] hope to solve a problem they cannot even formulate?"

In his classic 1917 book *On Growth and Form*, D'Arcy Thompson discovered that by drawing one species of fish on a rectangular grid and then deforming the grid by what we would now call a smooth coordinate transformation, he could obtain remarkably close likenesses of quite different, though closely related, species. An example is shown in Figure 2.14 where, starting with the species in the upper left-hand corner, we obtain the other three species by different bendings of the original rectangular grid. In carrying out these

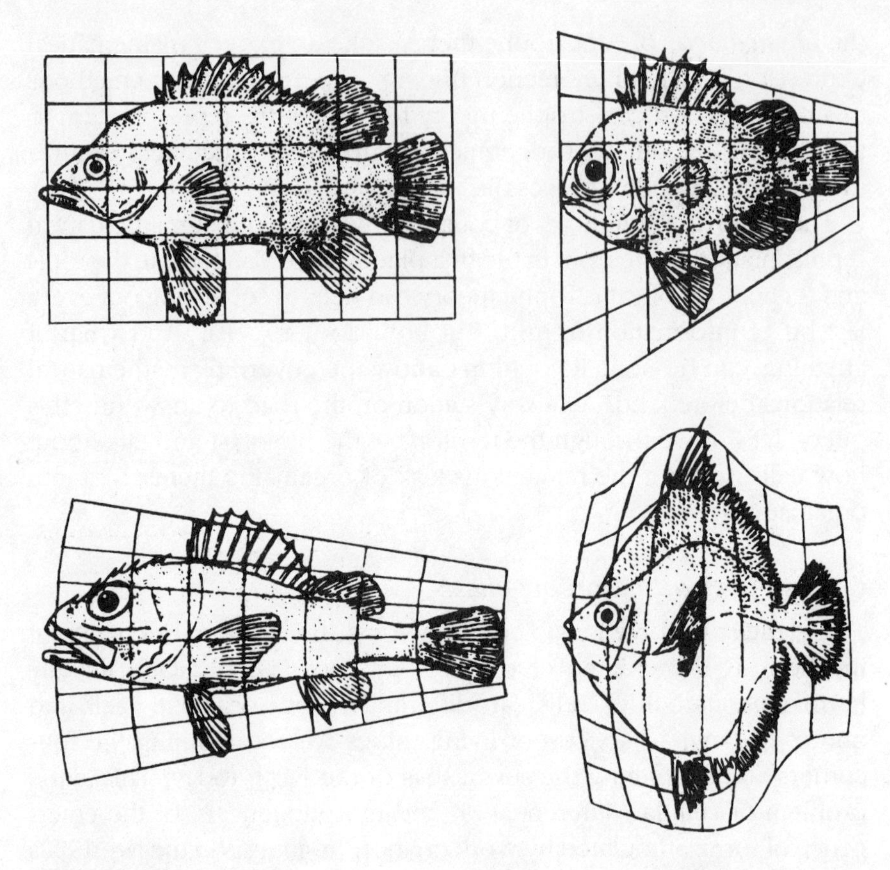

FIGURE 2.14. RELATED SPECIES OF FISH OBTAINED BY COORDINATE TRANSFORMATIONS

experiments, Thompson found similar transformations linking such diverse structures as the skulls of primates and the carapaces of crabs. These studies strongly suggest that topology is basic to the overall plan of an organism. The organism's *specific* geometrical form is filled in later, thus making the organism more amenable to modification during the course of evolution.

So the big question here is how the basic topology of the organism is specified. Morphogenesis doesn't take place simply by cells of the right type forming up in appropriate positions in the growing embryo. In fact, early on in the embryonic process, cells are not usually determined at all. Rather, most cells are capable of ultimately developing into one of a number of different cellular types for some period of time after their initial formation. So what is it that causes a nondescript, general-purpose type of cell to suddenly become a liver cell or a muscle cell or a brain cell? And once its fate is decided, how does the cell know the correct spatiotemporal pattern to follow in order to take its proper place in the final adult organism? In a nutshell, these questions constitute the mystery of morphogenesis.

Nobody really knows how the fate of an individual cell is decided. What is clear, however, is that neighboring cells interact and that there are chemical gradients within the growing embryo. One of the first researchers to propose an explicit mathematical model to explain development based on these chemical gradients was, surprisingly enough, the computer scientist Alan Turing. In 1952, Turing proposed a mathematical model involving the processes of reaction and diffusion of various unspecified chemical compounds that he termed *morphogens*. Many investigators since that time have developed Turing's idea to a high degree of mathematical sophistication, leading to the current conventional wisdom in developmental biology that the ultimate fate of each cell is decided by the concentrations of various morphogens. These concentrations, in turn, are assumed to be determined by a dynamical system describing how the morphogens interact and change their levels and types over the course of time.

Adopting Turing's basic idea, it's reasonable to suppose that there are time-dependent changes going on in the cells. To keep things as simple as possible, assume that the fate of each cell is decided by a single morphogen whose concentration is the fixed-point attractor of a dynamical system. These attractors will, of course, depend on hundreds, or even thousands, of biochemical variables, which in turn generate the postulated morphogen. We can think of these biochemical quantities as being internal variables, inaccessible to observation. So the only observed output of the system is the single quantity, the morphogen concentration. Since there are only four independent directions in space and time in which we can look at the developing organism, there are four independent input variables.

We now find ourselves in a familiar situation: a single output variable (the morphogen concentration) depending on four input variables, the spatial and temporal directions in which we can observe the developing embryo. We can think of these inputs as being knobs to twist, each setting specifying a particular location in space and time at which we observe the organism. Generally, the morphogen concentration, hence the observed properties of the organism, will vary smoothly as we slowly and continuously twist the knobs. However, for some knob settings, more than one stable morphogen concentration will be possible. This results in the formation of "frontiers" between different types of tissues and between these tissues and the spatial region outside the organism (i.e., the outside world). Thus, we conjecture along with Thom that in the local vicinity of a particular point in space and time, the physical form of the developing organism will be determined by one of the seven elementary catastrophes listed earlier in Table 2.1.

It turns out that each of the "magnificent seven" catastrophe types listed in Table 2.1 can be given both a spatial and a temporal interpretation, depending upon whether we interpret the input variable(s) as space or time. Table 2.2 lists these possibilities.

As an illustration of how to make use of these interpretations, consider Figure 2.15, which shows a section of the parabolic umbilic catastrophe. The spatial interpretation of this catastrophe (from Table 2.2) suggests the formation of a mouth, which is indeed a good image of what seems to be emerging as we take different sections of this catastrophe surface. Here the figure shows a striking similarity between the section of the parabolic umbilic and a bird's beak. Many other examples of this sort can be found in the references cited in the To Dig Deeper section for this chapter.

Developmental biology is an area somewhere between physics and philosophy when it comes to being able to write down an explicit mathematical representation for the underlying dynamical processes. We know *something* about these morphological processes. But we don't know nearly as much as we know about phenomena like the

TABLE 2.2.
SPATIAL AND TEMPORAL INTERPRETATIONS OF CATASTROPHES

Catastrophe	*Spatial Interpretation*	*Temporal Interpretation*
Fold	Boundary	Beginning (ending)
Cusp	Pleat; fault	Separating (uniting); changing
Swallowtail	Split; furrow	Splitting; tearing
Butterfly	Pocket	Giving (receiving); filling (emptying)
Hyperbolic umbilic	Wave crest; arch	Collapsing; engulfing
Elliptic umbilic	Spike; hair	Drilling
Parabolic umbilic	Mouth	Opening (closing); ejecting

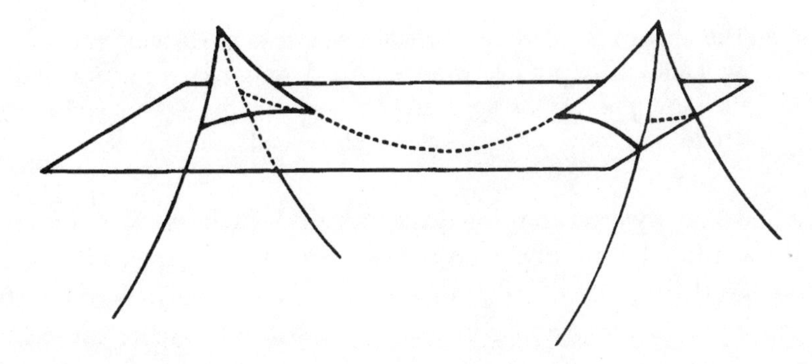

FIGURE 2.15. A SECTION OF THE PARABOLIC UMBILIC CATASTROPHE

light caustic studied earlier. Still, for intrepid explorers, ignorance of mechanism is never a deterrent to a search for knowledge. So let's push on even deeper into the twilight zone of speculation, considering how one might use catastrophe theory as a vehicle for modeling something as vaguely specified as the death of a civilization.

The Collapse of Civilizations

In studies of the collapse of ancient civilizations such as the Classic Maya, Mycenaean and Hittite, British archaeologist Colin Renfrew has identified the following characteristic features in the collapse phase:

- Disappearance of the traditional elite class
- Collapse of the central administrative organization
- Collapse of the centralized economy
- Settlement shift and population decline

During the aftermath period, we observe a transition to a lower level of sociopolitical integration and the development of a romantic Dark Age myth. In addition, such collapses display the following temporal features:

- The collapse may take on the order of one hundred years for completion, although in the provinces of an empire the withdrawal of central authority can occur more rapidly.
- Dislocations are more evident early in the collapse period and show up as human conflicts like wars and destruction.
- Border maintenance declines during the period so that outside pressures can be seen in the historical record.

- The growth of many variables like population, exchange and agricultural activity follow an S-shaped logistic growth curve (see the description of Cesaré Marchetti's work in the opening chapter).
- There is no obvious single cause of the collapse.

As a first step toward capturing this system in formal terms, let's follow Renfrew and take the observed output variable for such collapses to be D, the degree of centrality or control of the governing authority. As input parameters, we use investment in the charismatic authority I and net marginality for the rural population M. Here charismatic authority is not a measure of popularity or belief. Rather, it is the energy the society assigns to cultural devices used to promote allegiance to the central authority. The second input variable relates to the economic balance for the rural population. On the assets side are the fruits of their agricultural or craft activities. Liabilities include the material contributions, in goods or labor, required of them by the central authority as the price of citizenship and to escape punishment. Thus, the net marginality is the degree to which the assets outweigh the liabilities.

Using these quantities, together with the very big assumption that the underlying dynamics linking these inputs and output is smooth and has only fixed points as its attractors, the Classification Theorem assures us that the right geometry for this situation must be the cusp. Using this result, Renfrew arrived at the geometrical picture for the decline and fall of the society shown in Figure 2.16.

We can use this diagram to plot the course of a typical system's collapse like that of the Mayan or Mycenaean civilizations. Following Renfrew's account, the story starts at point 1. Marginality is low, implying that investment is in a charismatic authority. Furthermore, the degree of centrality is low, so this is a prosperous noncentered society, which may well be egalitarian. If I, the commitment to the charismatic authority, increases through points 2 and 3, the degree of centrality also increases. Such an increase might come about, for example, as the response to an external threat. Thus, the state develops. But marginality may now increase as well since it is no longer easy for the rural population to increase the per capita yield in order to make the additional contributions required by the central administration. As population increases, fertility decreases and/or the

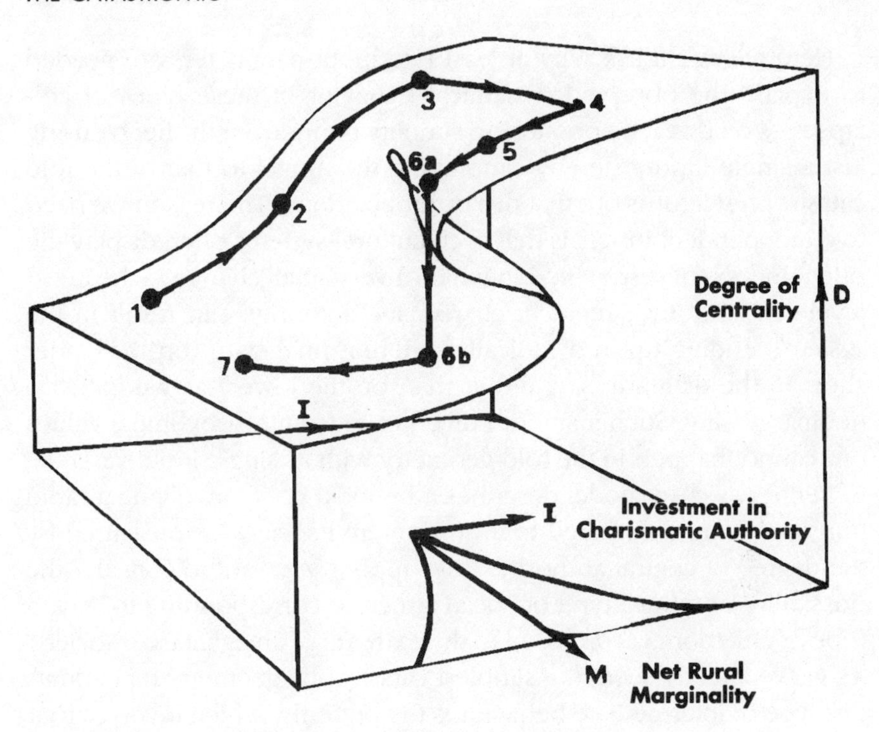

FIGURE 2.16. CUSP CATASTROPHE MODEL OF SYSTEM COLLAPSE

tax burden is augmented; thus marginality increases, leading to a corresponding increase in *I* to point 4.

Now the system is under stress, with high marginality, and *I* decreases slightly to point 5. In addition, there is now a lower value of *D* for which efficiency is also a maximum. But it's not until point 6a, when the local maximum vanishes altogether, that the value of *D* changes suddenly (point 6b). This very rapid drop involving the collapse of the central government brings in its wake many other changes. In particular, the central personnel, no longer exercising control or imbued with the charisma that accompanies it, can no longer command *I*, the investment in charismatic authority. The sharp collapse of *D* (the rupture of centralized control) is followed by the rapid but slower diminution of *I* to point 7. The administrative population, with its specialist officials and craftsmen, either dies, emigrates or returns to rural cultivation, and marginality is reduced (to point 1).

Here we might ask why at least two input parameters are needed to explain the observed dynamical behavior of these types of collapses. Wouldn't it be possible to simplify things even further by using just a single input, thereby simplifying the model to that of the fold catastrophe? It turns out that this cannot be done. The reason we need two independent inputs is that such cultural systems often display the phenomenon of *divergence*, in which a very small change in the initial level of I, the investment in charismatic authority, can result in the system's ending up in a radically different final state (on the upper sheet in the diagram, say, rather than on the lower) as we increase net marginality. Such a smooth divergence from nearby initial values of I cannot happen in the fold geometry with its single input variable.

Renfrew's cusp model describes a bimodal polity involving a rapid transition from a centered to a noncentered society, as measured by the degree of central authority. But suppose we want to consider the possibility of a third type of social structure corresponding to, say, a tribe or chiefdom, as opposed to the extreme of an egalitarian society. As we've already seen, the simplest catastrophe geometry that admits this type of intermediate behavior is the butterfly, which involves four input variables. So we're going to have to add two additional factors to the quantities I and M used above. Let's take T and K as these two new quantities that determine whether or not a tribal structure is possible. The problem is what kind of interpretation to give to these new inputs. What do these purely mathematical necessities really *mean* in the context of chiefdom formation? Looking at the butterfly geometry of the situation will help us answer this question. This geometry is shown in Figure 2.17.

Translating this picture into words, only bimodal behavior is possible and local chiefdoms cannot arise when the butterfly factor K is negative. But when K is positive, trimodal behavior is the rule. Thus, it's reasonable to propose that K be regarded as a measure of the extent to which relations and rank in the society are determined by kinship (i.e., the relations are the result of birth or marriage). In other words, positive values of K mean that kinship is very important in determining societal rank, while negative values mean antinepotism is the rule. So what makes a society a chiefdom rather than a state is the extent to which relations within the society are still determined by kinship.

The bias factor T determines mathematically which of the two "wings" of the butterfly is the more dominant. In one case, the society

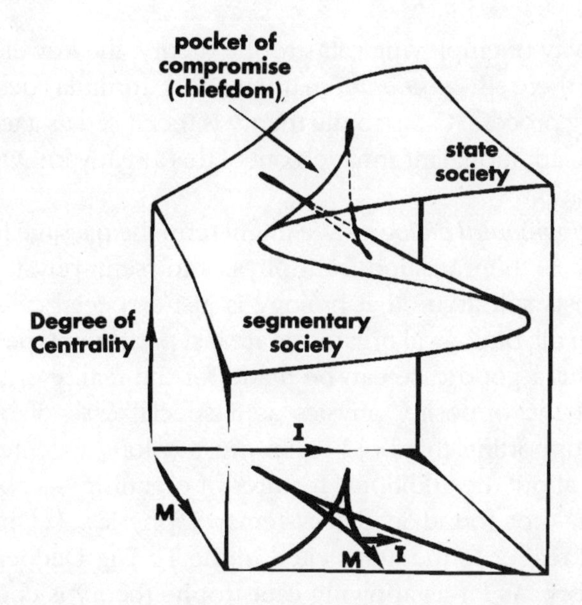

FIGURE 2.17. FORMATION OF CHIEFDOMS AS SEEN BY THE BUTTERFLY CATASTROPHE

is more statist-oriented, while if the other wing is dominant there is a more pronounced tendency toward societal fragmentation. So one plausible way to interpret *T* is as the intensity of the external threat to the system. If the threat is low, fragmentation will dominate; if it's high, the society will tend to organize itself around a central authority.

In philosophy, the term *metaphysics* refers to the study of questions that transcend the merely natural, issues such as whether every event has cause and what things are genuinely real. With our last example on the collapse of early state societies, we have clearly moved from the realm of the physical to the metaphysical insofar as our use of catastrophe theory goes. And it is exactly this transition that ruffles the feathers of those critics of how the theory is sometimes applied. As a prelude to the consideration of their arguments, let's stand back and take a bird's-eye view of the three examples in this section.

- *Bending beams*—Here well-tested principles and laws of classical physics give an explicit mathematical expression for how the beam behaves. By applying the methodology of catastrophe theory to this mathematical representation, we were able to determine directly the canonical geometry that must govern the beam's behavior. In this physical way of employing catastrophe theory, the key element is the fact that there is a *known* mathematical formula describing the underlying process. Catastrophe theory is then used as a tool by which to squeeze additional information out of this already known and tested representation.

• *Developmental biology*—We might term the passage from physics to biology as the transition from physics to "semi-physics." Die-hard reductionists still argue that biology is just a special case of physics since, after all, biological organisms are just particular types of physical objects. But a good case can be made for the reality of the situation being just the opposite: physics as a special case of biology. The reasons supporting this bold claim are too long to enter into here, revolving about the additional features of organisms—replication and self-repair—not found in the systems of physics. Let me refer the interested reader to the items cited in the To Dig Deeper section for the full story. As far as applying catastrophe theory is concerned, the central ingredient is that in biology we generally do not have the luxury of an explicit mathematical representation for the process under investigation, unlike most situations in physics. But this does not mean that nothing is known.

For example, in studying blood flow through the vascular system, we have a very complete set of equations describing the blood movement, the well-known Navier-Stokes equations of hydrodynamic flow. In this case we can use catastrophe theory in the physical way, exactly as it was used for the case of the bending beam. In fact, for such situations it's difficult to draw a meaningful distinction between biology and physics, at least insofar as the formal modeling activity is concerned. But when it comes to a problem like morphogenesis, matters take a far different turn.

In our example above, we saw that there was a mathematical formalism—Turing's reaction-diffusion equations—governing the behavior of a hypothetical morphogen postulated to account for the developmental pattern of a group of cells. The problem here is that, first of all, no such morphogen has ever been indisputably identified in the lab. So it's hard to argue that Turing's equations are anything like well tested by experiment. Moreover, even if there were a physical quantity we could confidently label a morphogen, the gap in the causal chain between any such morphogen and actual cellular behavior is enormous. There is no mathematical way known today to bridge this chasm. And, sad to say, this is the situation for a large number of problems of interest in modern biology.

So when it comes to applying catastrophe theory in the biological sciences, we must take refuge in a kind of semiphysical way, in which we assume that whatever mathematical relations govern the situation, they satisfy all the requirements needed in order to invoke the Classification Theorem. Once this hurdle has been surmounted, it's smooth sailing thereafter, as we then need only identify appropriate input and output quantities and look up the corresponding standard geometry on Thom's list. But note carefully that in this approach we have at least some mathematical basis upon which to ground our arguments. If we're lucky, we have a clear-cut representation like the Navier-Stokes equations; if not, then we have only an indirect connection to the phenomena at hand, as with Turing's equations in morphogenesis.

• *Collapse of civilizations*—Turning to the social and behavioral arenas, it's difficult to imagine anyone claiming with a straight face that they have an explicit mathematical representation for how individuals, or even societies, act. In fact, many sociologists and anthropologists of a somewhat traditional turn of mind would probably argue that there is no such representation. Or if there is, it's beyond the powers of the human mind to find it. Either way we're groping in the dark insofar as painting a usable mathematical picture of human behavior. So how can catastrophe theory be applied in such situations?

The basic approach is to use what has come to be termed the *metaphysical way* of catastrophe theory. One begins by singling out a handful of variables, calling them inputs. These quantities, in turn, are *postulated* to give rise to observable changes in another handful of quantities that we term outputs. And to keep things within the confines of elementary catastrophe theory, we further *posit* that there are no more than six inputs and two outputs. Next we *assume* that whatever the mathematical relationship linking the inputs to the outputs, the analytic structure of that relationship satisfies the technical conditions required by the Classification Theorem. In broad terms, these conditions amount to the requirement that the system's outputs be fixed-point attractors of some smooth dynamical system. Finally, we *presuppose* that the coordinate systems used to measure both the inputs and the outputs happen to coincide with the coordinate systems that lead to the standard catastrophe geometry governing a system with the number of inputs and outputs we have chosen. If not, we have to make a change of variables to transform the physically based coordinates in which we express the observations into the mathematically based variables that generate the standard geometries. The mathematics of catastrophe theory assures us that such a transformation exists—but gives no help in finding it.

Now focus your attention on the italicized words in the preceding paragraph. Each one of these words represents an out-of-the-blue assumption that we must be ready to swallow if we want to appeal to the Classification Theorem to single out a particular geometry describing our problem. It's at this point that physics gives way to metaphysics. It's also at this point that the critics of catastrophe theory come crawling out from under their rocks to level their multifarious attacks on what they see as the spurious, if not downright dangerous, applications claimed by the theory's proponents. So without further ado, let's turn the podium over to these critics and answer the question raised at the beginning of the chapter: how can a mere mathematical theory of discontinuous processes lead to the kind of bitter intellectual dispute that's usually found only in the arts and humanities?

THE THEATER OF THE ABSURD

Happily, the controversy over catastrophe theory is a lot easier to understand than the theory itself, once we recognize that the controversy really separates into four quite distinct arguments about: (1) the theory's mathematical and philosophical foundations, (2) the assumptions necessary to apply the theory, (3) the details of particular applications, and (4) the scientific attitudes, styles and even personalities of the theory's supporters and opponents. We have already considered the second point within the context of the physical and metaphysical ways of catastrophe theory, so the balance of our discussion in this section will focus on the remaining elements of the debate.

The principal points raised against catastrophe theory from a mathematical point of view are that the theory is inherently local, telling us only about what's happening in a small region of the output space, and that it applies only to a very restricted class of dynamical processes. Specifically, it applies only to dynamical systems having fixed points as their attractors.

The problem with locality is that the Classification Theorem doesn't tell us what the system's behavior is like far from a critical point. Furthermore, since the standard geometries like the cusp and butterfly admit stretching and bending of both inputs and outputs without changing the topological character of the critical points, it takes a certain act of faith to associate a mathematical jump in the space of attractors with an observable discontinuity in some real-world process. The difficulty is that the mathematical jump may be too small to be seen—or even regarded—as a jump in the real-world process the catastrophe is supposed to represent. Developing the tools needed to create a global theory of catastrophes that would justify this act of faith is a big conceptual and mathematical problem, one going far beyond the resolution of mere technical obstacles in the existing theory.

As for the restriction to systems with fixed-point attractors, it turns out there are many natural processes whose long-run behavior is not the fixed point of a dynamical process. Hence, such systems cannot be governed by the type of dynamics needed to apply elementary catastrophe theory. For instance, many biological processes like the human heartbeat and the flowering of plants involve steady-state behaviors that are periodic. And as we will see in the next chapter, an even larger class of systems seem to settle into attractors that are "strange." In both cases, catastrophe analysis is inapplicable unless the problem has some kind of special structure allowing us to reduce the behavior of interest to the right kind of dynamic.

Despite these incontrovertible mathematical facts, it is still the case that many interesting processes *do* fall within the purview of elementary catastrophe theory. Furthermore, it's often possible to closely approximate many systems of the "wrong type" by those having fixed points as attractors. But these mathematical and philosophical difficulties are just the antipasto to the *real* catastrophe theory controversy. The main course revolves around the way the theory has been applied and the perceived motivations of its main practitioners, especially Christopher Zeeman.

In our discussion of societal collapse, we saw that a number of assumptions have to be accepted in order to use catastrophe theory in situations where we do not know the explicit mathematical form of the underlying dynamics. We termed this the metaphysical way of catastrophe theory. And it's exactly this transition from physics to metaphysics that constitutes one of the main lines of attack against the application of catastrophe theory. In the words of Hector Sussman and Raphael Zahler, two of the leading anticatastrophists, "catastrophe theory is one of many attempts that have been made to deduce the world by thought alone . . . an appealing dream for mathematicians, but a dream that cannot come true." This sentiment is echoed by dynamical system theorist John Guckenheimer, who wrote in a review of Thom's book for the *Bulletin of the American Mathematical Society*, "they [Thom and Zeeman] show a real reluctance to get their hands dirty with the scientific details of the applications." But Guckenheimer concluded his review by saying that Thom might even be too cautious about the impact of his theory on biology.

There are several aspects to these reservations about the applications of catastrophe theory, some centering on Thom's stated assumption that a good model for a physical process must be structurally stable, while others zero in on the metaphysical assumptions that the system has only fixed points as its attractors and that its behavior depends on only a small number of input parameters. Still another line of attack has been the claim that many of the published applications of catastrophe theory are just plain wrong, in the sense that they give predictions that are flatly contradicted by observational evidence.

Concerning the structural stability objection, John Guckenheimer has noted that "there are very reasonable models for occurrences in the real world that simply are not structurally stable, or even qualitatively predictable." In other words, not every good (i.e., useful) model need necessarily be structurally stable. We will see evidence in support of this claim in the next chapter when we look at processes governed by chaotic dynamics. Such systems are often manifestly unstable, small shifts in the system parameters giving rise to transitions from simple steady-state motions like laminar flow in a fluid to complicated, inherently unpredictable behavior like fully developed turbulence. The theory of elementary catastrophes as it currently exists cannot account for these kinds of behaviors, and it remains an open question as to whether a suitable theory of "generalized" catastrophes can be developed to embrace such unstable processes.

As for catastrophe-theoretic descriptions giving rise to bad models, one of the most flagrant examples of this sort is Zeeman's cusp model of the boom-and-bust behavior of prices on speculative markets. Here it suffices just to note that the model implies that in a purely speculative market involving no investors who trade on the basis of economic fundamentals, there can never be a crash or a boom. If the only information investors use to make decisions is past price data, then there can never be a discontinuous change in prices. On the surface such a conclusion seems patent nonsense. Real markets always experience booms and busts, and any model implying otherwise must be taken with several shakers full of salt. Other objections along similar lines have been leveled at many of Zeeman's models of things like prison riots and the aggressive behavior of animals. The degree to which these complaints hold water will ultimately turn on whether or not these preliminary models can be revised to remove the offending pieces without throwing out the catastrophe-theoretic baby with the bathwater. But judging from the lavish praise and the howls of outrage about catastrophe theory quoted at the beginning of this chapter, it appears that the most viscerally interesting aspect of the controversy is not mathematical at all, but psychosocial. So let's consider the human side of the debate.

Public images notwithstanding, one should never forget that science, like literature, music, dance and all the other arts, is a human activity. In fact, there's ample evidence to suggest that scientists are far from immune to the emotional outbursts and petty jealousies that are endemic and taken for granted in these more humanistic areas of intellectual pursuit. Many years of empirical evidence attest to the fact that mathematicians take a backseat to no one when it comes to hypersensitivity to the folkways and mores of their profession, especially when one of their number receives attention beyond the cloistered walls of academia. The catastrophe theory controversy is one of the best examples I can think of in support of this thesis.

In its issue of January 19, 1976, *Newsweek* magazine ran a full-page article on catastrophe theory, the first story on mathematics they had published in over seven years. In this article the theory is described in terms rosy enough to emit heat, suggesting that Thom's ideas about discontinuous phenomena represent the most significant advance in applied mathematics since Newton's invention of the calculus. As we have come to expect whenever a laborer in the vineyards of mathematics receives even a smidgen of attention beyond the bounds of what the mathematical community feels is right and proper, the naysayers came out in force. In this case, the charge was led by the aforementioned Sussman and Zahler, aided and abetted by a number of prominent colleagues. In a 1977 article by Gina Kolata in *Science* magazine titled "The Emperor Has No Clothes," the battle was joined in earnest, with several quotes on the demerits of catastrophe theory being uttered by prominent mathematicians like Stephen Smale of Berkeley and Joseph Keller of Stanford. In Kolata's notorious article, Zeeman is described as a "publicist," and Thom is rightly quoted, but completely out of context, to the effect that in a world in which all concepts could be formulated mathematically, only the mathematician would have a right to be intelligent. What followed was a long bout of correspondence, pro and con, on the issues raised in the article, little of which had any bearing on anything other than the attention that Thom and Zeeman were receiving from the world outside science and, especially, outside mathematics.

In trying to summarize and evaluate the pluses and minuses of the many threads in this debate, it's difficult for an uninvolved bystander not to wonder, Why all the fuss? As biologist Robert Rosen wisely counseled, "If an individual scientist finds such concepts uncongenial, let him not use them. There is no reason why he should take their existence as a personal affront." This is my view as well. Catastrophe theory will probably survive these broadsides, in much the same way and for much the same reasons that Darwin's theory of natural selection survived the bitter attacks mounted against it. Both theories are essentially explanatory rather than predictive, thereby failing to provide those who hunger for precise quantitative predictions with the kind of numerology that has come to be synonymous with *science*. But as René Thom so poignantly points out, "At a time when so many scholars in the world are calculating, is it not desirable that some, who can, dream?"

THREE

THE CHAOTIC

Intuition: Deterministic rules of behavior give rise to completely predictable events.

Chance is perhaps the pseudonym of God when He did not want to sign.

—Anatole France

The world is either the effect of cause or of chance. If the latter . . . it is a regular and beautiful structure.

—Marcus Aurelius

Everything existing in the Universe is the fruit of chance and necessity.

—Democritos

Expecting the Unexpected

The University of Bologna is reputed to be the oldest university in the Western world, having been founded sometime in the 11th century. And by the latter part of the Middle Ages, the science faculty of this venerable institution had a chair in astrology. To the modern eye this fact seems incongruous, to say the least, as there's probably not a single facet of the occult that the modern scientist condemns more vociferously than astrology. An obvious question then is why astrology was regarded as the height of scientific respectability a few centuries ago, but is now thought of as the archetypical pseudoscience.

Some years ago, the vagaries of fate threw me into extended contact with a mathematician and writer who thought of himself as a "chronotopologist." Having studied a bit of topology myself, I was more than a little curious to know something about the topology of time, especially the way in which it might serve to complement the traditional topology of space that I had learned about at university. I became even more curious when it turned out that my friend was doing nothing more nor less than practicing astrology under a New Age kind of label. What a golden opportunity, I thought, to hear straight from the horse's mouth about how one could simultaneously swear allegiance to both the tenets of classical astrology and the precepts of modern science.

The key to reconciling these apparently conflicting visions of reality is to accept the following hypotheses: (1) there are natural forces that really do influence human destinies, and (2) the nature of these forces cannot be accounted for by today's science. My astrological colleague then claimed that the traditional view of the planets as being the actual *source* of these mysterious forces affecting people's lives is completely mistaken. The planets exert no such influences, he argued. Rather they serve only as a convenient way to *measure* these fundamental forces unknown to modern science. And it is these deeper, more basic forces that actually do cause people to get rich, get sick, or get lucky.

A good analogy here is electricity flowing through a wire. By attaching a voltmeter to the wire, you can measure the flow of current, which will be displayed as the position of the voltmeter's needle on a graduated scale. But that needle movement is not the force itself; it's only an *indicator* of a force—namely, the electrical potential existing in the wire. According to the dictates of chronotopology, so it is with the planets. Their positions are only a convenient way of measuring the vital "life force" that astrologers believe influences every one of us from birth to death. But if the planets don't directly influence our lives, then why do we use them as the measuring device? Why not use, say, the growth patterns and colors of the flowers in Central Park or the configuration of clouds outside your window?

Chronotopology argues that any of these other natural phenomena can in fact be used as an indicator of the crucial life forces so beloved by astrologers. We customarily use planetary positions only because these positions can be accurately calculated years, even decades or more, into the future. Thus, they serve as a reliable indicator of what the mysterious astrological forces will be doing and how they will be acting for as far into the future as we care to compute the planetary orbits. So the bottom line turns out to be that the popular image of astrology is all wrong. There is no special relationship between the positions of the planets and the puny affairs of us humans. Instead, the planets come into play simply because it's easier to calculate their orbits with greater precision and over longer time horizons than it is to easily and reliably predict the long-term future of any other natural process.

The point of this little story is to underscore the fact that the laws governing planetary motion are about the closest thing the human race has yet discovered to a sure thing. Measure the current positions of Mercury, Saturn, Venus or Mars, plug these measurements into Newton's laws of motion and out come extremely accurate predictions of where the planet will be tomorrow, next month or next century. This is the example par excellence of what we mean when we talk about a system's behavior being "determined."

In the 18th century, following in the footsteps of Newton, Pierre-Simon de Laplace, physicist, mathematician and one of the founders of the theory of probability, made a bold assertion that is about as good a description as can be given of the perceived relationship between determinism and predictability:

> Given for one instant an intelligence which could comprehend all the forces by which nature is animated and the respective positions of the beings which compose it . . . nothing would be uncertain, and the future as the past would be present to its eyes.

But Laplace was wrong. Being deterministic and being predictable are just not the same thing at all.

A system is called *deterministic* when its future states are completely fixed by its current state and its rule of dynamical motion. So, for instance, if we know the current positions and velocities of the nine planets, then the laws of celestial mechanics fix uniquely the position and velocity of each planet for all future times—in principle, anyway. The simple Circle-10 system we saw in the last chapter is another good example of a deterministic system: once we specify the starting point on the circle, the itinerary is fixed thereafter by the rule telling us how to move from one point on the circle to the next. But the Circle-10 system also underscores the vitally important fact that deterministic behavior does not necessarily imply complete predictability. It's crucial for our understanding of chaotic phenomena to know why this should be the case. Basically, a chaotic system generates behavior giving the *appearance* of complete randomness by means of a purely deterministic rule. For a truly random process, on the other hand, there is no such fixed, deterministic rule. Rather, such a process arises from an inherently probabilistic rule. Our principal goal in this chapter is to see how to distinguish between these two cases. The first is what scientists call chaotic, while the second is what laymen have in mind when they use the very same term. Scientifically speaking, chaos is only the appearance of randomness, not the real thing.

RECIPES FOR RANDOMNESS

Judging by the number of cookbooks on display at my local bookstore, cooking must be about the most universal of human activities, rivaled only by making money, love, war or trouble. And what does every one of these cookbooks consist of? Basically, these books are collections of rules telling us how to create just about anything from aardvark stew to zucchini bread. For example, consider a recipe for making something nice and fattening like bread pudding. The recipe begins by specifying all the ingredients you must have on hand in order to make this delicacy. These items include things like milk, eggs, butter, salt and, of course, lots of dried bread. Speaking in computer jargon, these ingredients constitute the *input* to the bread-pudding system. The recipe then sets out the sequence of steps you must follow to actually make the pudding. Roughly speaking, each of these steps tells you to do something with either the raw ingredients or an intermediate product, assuring that if you don't make any mistakes along the way, the final step will take you to the object of your desire. Again in computer jargon, we can call the steps of the recipe a *program* for bread pudding.

So we see that there is no fundamental difference between a computer program and a cooking recipe, other than the fact that the recipe tells us how to manipulate physical matter while the computer program tells us how to manipulate information. Following the same line of reasoning, it should be clear that a computer program and a dynamical system are also abstractly equivalent objects. To use the kind of obfuscation that mathematicians revere, they are *isomorphic*. The initial state of the dynamical system corresponds to the input to the program, while the dynamical rule of motion (the vector field) corresponds to the instructions that make up the program. And, in fact, with these matchups in hand, we can consider a cooking recipe to be a particular kind of dynamical system. Let's pursue this isomorphism just a little bit further, both in its own right and as a way of sneaking up on our original question about the nature and origin of chaos.

Suppose you're making bread pudding and inadvertently put in a pinch too much salt. Chances are this departure from "design specs" won't do much harm to the final result, and you'll end up with something pretty close to a bona fide bread pudding. In the language of dynamical systems, the bread-pudding system is stable: a small change in the input to the system causes the output to wind up in an attractor close to that representing the perfect pudding. Of course, if you dump in a whole shaker full of salt the end result will bear little resemblance to anything we'd even charitably want to call bread pudding. This is what we mean when we say that the bread-pudding system is *locally* stable. Small changes to the input and/or to the processing rules result in small changes in the final product. But all bets are off if we make large changes. And this is not a situation confined just to the kitchen.

The Butterfly Effect

In the early 1960s, Ed Lorenz, a meteorologist from MIT, was experimenting with some computational models of the atmosphere. During the course of one of his computer experiments, he discovered what has now come to be called the *butterfly effect*. This terminology comes from the fact that in weather models like Lorenz's, a butterfly flapping its wings today in Brazil can jiggle the atmosphere so as to cause a snowstorm in Alaska tomorrow. Put more technically, the purely deterministic laws governing weather formation are unstable in the worst possible way. As a result, they allow minuscule changes at one location to percolate through the system so as to bring about major effects somewhere else.

So we have the following puzzle: The butterfly effect is a small change, yet it leads to a very large change in the final result. Unlike the bread-pudding system, the weather system is unstable—even locally. Why? What's the difference between the dynamics of the atmosphere and the dynamics of bread pudding? To see why a butterfly beating its wings in the Amazon is fundamentally different from putting a pinch too much salt into the bread pudding, let's return for a moment to the Circle-10 system.

For the Circle-10 system, two starting points closer together than we could ever hope to measure turn out to follow itineraries that end up very far apart. This is due to what we earlier termed *sensitivity to initial conditions*, a property that the Circle-10 dynamics has in spades. The root cause of this behavior is that when we multiply by 10 in the Circle-10 rule, we are "stretching" the circle by a factor of ten. As a result, nearby starting points get pushed far apart. But since the circle is a bounded region, we can't stretch it everywhere and still remain on the circle. So after the local stretching, we have to fold the stretched-out circle back on itself in order to make it fit on top of the original circle. Expressed algebraically, this folding operation takes place in the Circle-10 rule when we drop the number appearing on the left-hand side of the decimal point after performing the multiplication by 10. Figure 3.1 gives a stylized view of the overall stretching-and-folding process.

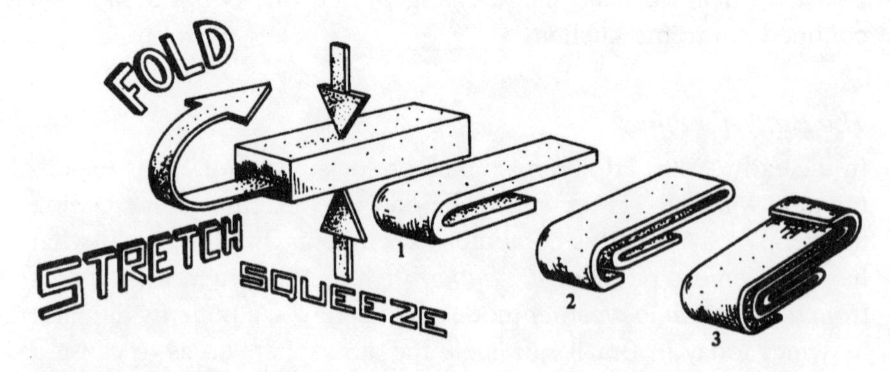

FIGURE 3.1. STRETCHING AND FOLDING

The stretching and folding operations are complementary, in the sense that the first separates points while the second tries to bring them back together again—but with new neighbors. By way of contrast with both the Circle-10 and the weather dynamics, local stretching is not the dominant effect in the bread-pudding recipe (i.e., its dynamics). In fact, there's no stretching at all in the bread-pudding rule. As a result, the sequence of states leading from the proper initial state (the correct ingredients) to the final pudding (the attractor) each remain close to the states on a path starting at a nearby initial state (e.g., one having a pinch too much salt). So the two itineraries each end up in the same or a nearby attractor, one that we're prepared to honor with the label *bread pudding*.

As a small historical aside, this idea of stretching and folding as the characteristic essence of chaotic processes is reputed to have been discovered in the mid 1970s by German theoretical chemist Otto Rössler of the University of Tübingen. As the story goes, one day Rössler was watching a saltwater taffy-pulling machine in Salt Lake City (where else?). What he saw through the shop window is shown in Figure 3.2. Apparently the machine was engaged in pulling a type of taffy containing raisins. While watching this process, Rössler

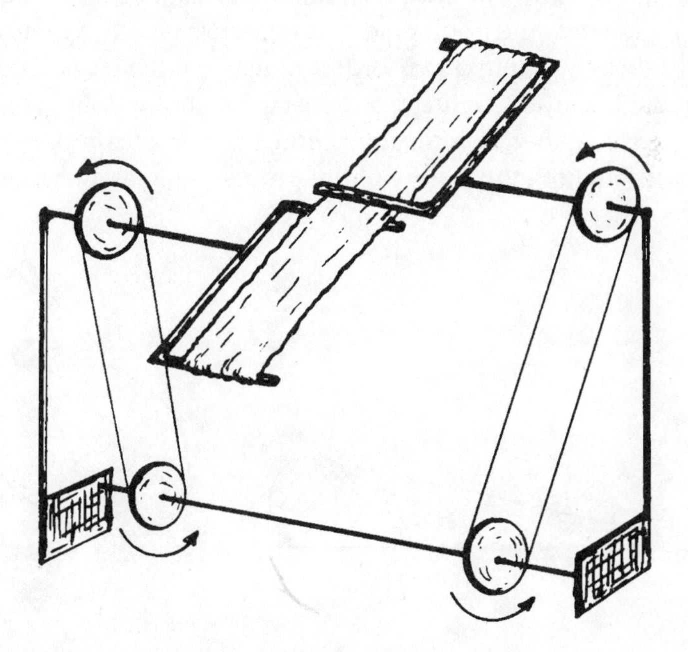

FIGURE 3.2. A SALTWATER TAFFY-PULLING MACHINE

wondered about the paths followed by two raisins initially thrown into the batch of taffy at about the same location. His contemplation of this question led to a scientific breakthrough of major proportions, which just goes to show that asking a childishly simple question doesn't always lead to a childishly simple answer—if it's the right question!

The preceding arguments show that for the bread-pudding system there is no sensitivity to initial conditions. Thus we can feel confident that a small change in the ingredients won't have much effect on the final pudding. It's a good thing, too, since it doesn't take much imagination to see what a mess we'd be in if cooking recipes were chaotically unstable like the Circle-10 system.

Sensitivity and Stability

The distinction between sensitivity and insensitivity is illustrated in Figure 3.3, where the first part of the figure shows the origin of the kind of sensitivity seen in the Circle-10 system. Here we see how it's possible to make the points of the real number line match those of a circle in a one-to-one manner. But note how the points far out on either end of the line (the *really* big numbers) all get compressed into a small region near the north pole. Thus, points that are close together on the circle may end up many orders of magnitude apart on the line. The second half of the figure represents the sort of stability we see in cooking recipes, in which nearby starting points remain pretty close together throughout the history of the process, ultimately winding up in the same attractor.

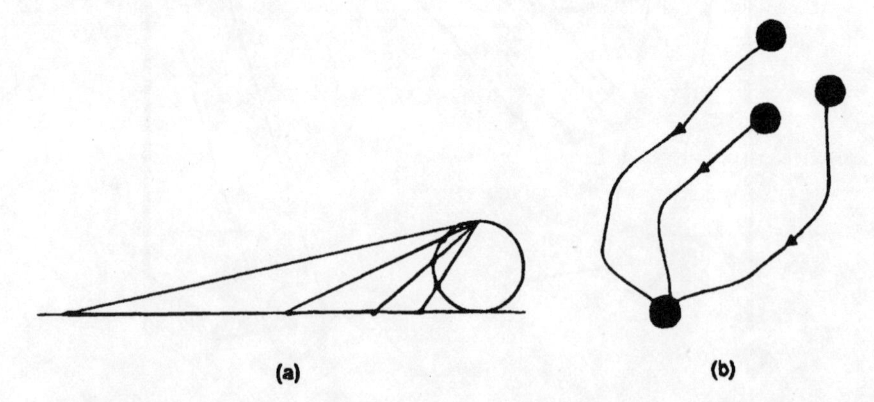

FIGURE 3.3. (A) SENSITIVITY TO INITIAL CONDITIONS; (B) INSENSITIVITY TO INITIAL CONDITIONS

Interesting and important as stable systems are for everyday life with its appeals to common sense and folksy rules of thumb, the majority of events we read about in the daily newspaper—lotteries, sporting events, stock price movements, the outbreak of wars—are the end result of an unstable process. When it comes to the kinds of things that really shake people up, it's butterflies that count, not bread pudding. So let's look a bit more closely at this kind of deterministic randomness in order to hammer home the point that chaos and common sense go together like oil and water.

Back in the early days of computers, one of the more popular methods for generating a sequence of random numbers was to employ the following scheme:

1. Choose a starting number between 0 and 1.
2. Multiply the starting number by 4 ("stretch" it).
3. Subtract 4 times the square of the starting number from the quantity obtained in step 2 ("fold" the interval between 0 and 1 back on itself in order to keep the final result in the same range).

Given a starting number between 0 and 1, we can use this procedure—often termed the *logistic rule* (see Chapter One)—to generate a sequence of numbers that to all appearances is completely random. For example, in such a sequence each of the ten digits 0 through 9 appears with equal frequency and the statistical correlation between groups of digits is zero. Note, however, that the members of this sequence are specified in a completely deterministic way by the starting number. So the sequence is certainly not random in the everyday sense of being unpredictable; once we know the starting number and the rule for calculating an element of the sequence from its predecessor, we can predict with complete confidence what every element in the sequence will be—providing we can pin down the starting number *exactly* and carry out the computational steps without error of any kind. Nevertheless, since such a sequence satisfies all the traditional statistical tests for randomness, it is called a *pseudorandom* sequence.

Insect Population Dynamics

In 1976, the prestigious British scientific weekly *Nature* published an article by Robert May, a mathematical ecologist now at Oxford

University, that served as one of the sparks touching off the explosion of current interest in chaotic dynamics. And what was the theoretical centerpiece of May's pathbreaking article? None other than a simple formula used to represent fluctuations in the population level of an insect colony, a setting in which the rule is generally labeled the *logistic equation*. In fact, May's logistic equation for insect population fluctuations turns out to be essentially the same rule as that sketched above for generating pseudorandom numbers.

Let's look at the logistic rule from an ecologist's point of view. First, we denote the insect population level at generation t by the symbol x_t. Assuming that when there is no competition for resources each adult produces four offspring, the population in the next generation will then be $4 \times x_t$. Now we scale the population so that levels beyond 1 signify a population explosion, while those less than 0 denote extinction. Then the quantity $1 - x_t$ represents a feedback on birth rates from effects due to overcrowding (i.e., competition for scarce resources). When we put the birth and overcrowding effects together into a single expression, what pops out is the formula for randomness we have called the logistic equation.

Figure 3.4(a) shows a hypothetical insect population generated in a computer experiment using the logistic rule, while part (b) of the figure shows actual data on the daily population fluctuation of blowflies obtained by A. J. Nicolson in a classic experiment carried out some decades ago. Even a casual glance at the figure suggests a strong connection between the mathematical randomness shown by the logistic law of growth and the unpredictable ups and downs of real insect populations.

Watching nature red in tooth and claw is certainly one way to see counterintuitive, chaotic behavior rear its ugly head. Looking at the clouds passing by outside your window is another.

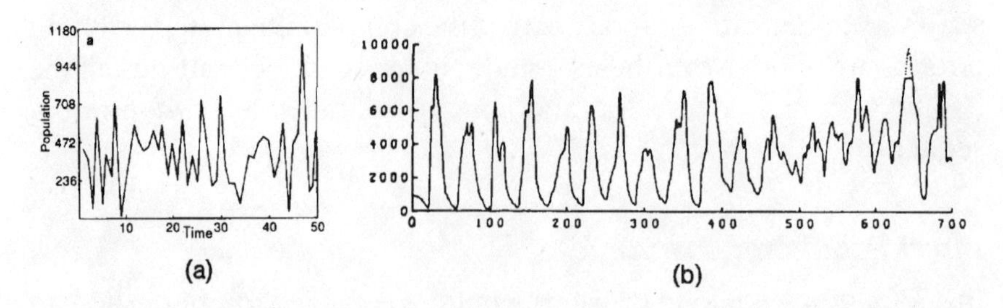

FIGURE 3.4. (A) POPULATION HISTORY FROM COMPUTER MODEL; (B) BLOWFLY POPULATION DATA

Atmospheric Dynamics

Many of the world's weather patterns are dictated by the position of the jet stream, at least in the mid-latitude regions where most of us work and play. Figure 3.5 displays the position of the jet stream over Europe and the mid-Atlantic in two quite different weather regimes. The first results in somewhat unsettled weather over the U.K., while the second may lead to sunny, warm weather in the summer and gray, possibly very cold weather in winter. To accurately predict these types of weather regimes is one of the principal tasks of the meteorologist.

In an attempt to model the real atmosphere, which involves something on the order of a million variables, Edward Lorenz developed a "toy" atmosphere represented by just three quantities: (1) the intensity of air movement, (2) the temperature difference between ascending and descending air currents, and (3) the temperature gradient profile between the top and bottom of the atmosphere. Let's label these three variables x, y and z, respectively. So in this setup, if x and y are both positive or both negative it means that warm air is rising and cold air is falling. Moreover, a positive value of z means that the greatest temperature gradient occurs near the boundaries of the atmosphere.

In Lorenz's model a weather state at a given moment in time is represented by a point in the three-dimensional abstract space whose points are determined by the variables x, y and z. Thus, the development of the weather over time can be thought of as the tracing out of a curve in this space. The set of all possible weather states forms what's known as the *Lorenz attractor* of the system. Its shape, shown in Figure 3.6, accounts for why Lorenz's work on measurement sensitivity and weather forecasting was originally labeled the *butterfly effect*.

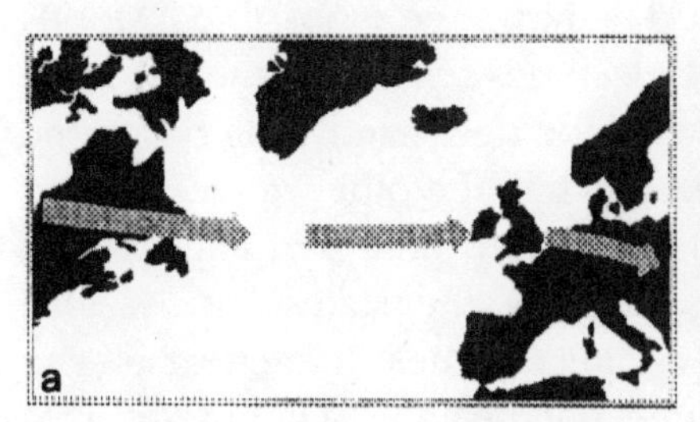

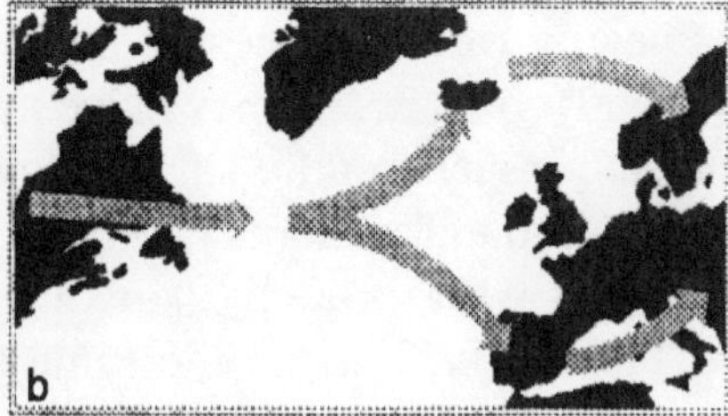

FIGURE 3.5. (A) UNSETTLED WEATHER PATTERN; (B) GOOD SUMMER AND BAD WINTER WEATHER PATTERN

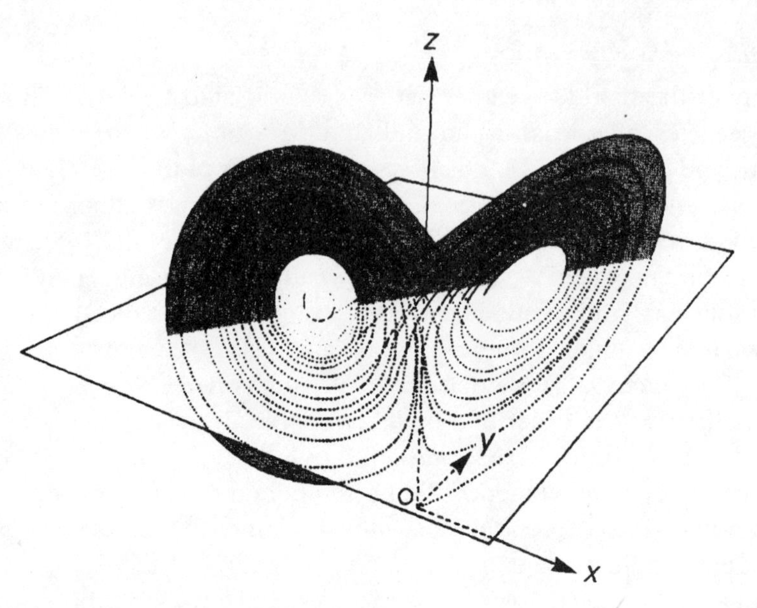

FIGURE 3.6. THE LORENZ ATTRACTOR

The first important point to note about the Lorenz attractor is that it has two separate "butterfly wings," which are abstract representations of two quite different weather regimes. For the sake of argument, let's assume that the left-hand wing represents the unsettled pattern of Figure 3.5(a), while the right-hand wing stands for the pattern of fair weather in summer and bad weather in winter depicted in Figure 3.5(b).

Now consider two points that are very close to each other on the left-hand wing. These points represent nearly identical weather states in the regime characterized by unsettled conditions over the British Isles. Let's try to follow what happens to these instantaneous states of the weather as time unfolds. Figure 3.7 shows three possible histories of nearby weather states: (a) both trajectories remain on the left-hand wing, (b) both trajectories move toward the right-hand wing or (c) one trajectory stays on the left-hand wing, while the other moves toward the right-hand pattern. It's important to note here that each of the three cases (a) through (c) involves initial weather states that are close to each other, but *not* the same two initial states in each case. The distance between the states is the same, but the initial states themselves differ in each case. It is this fact, coupled with the chaotic behavior

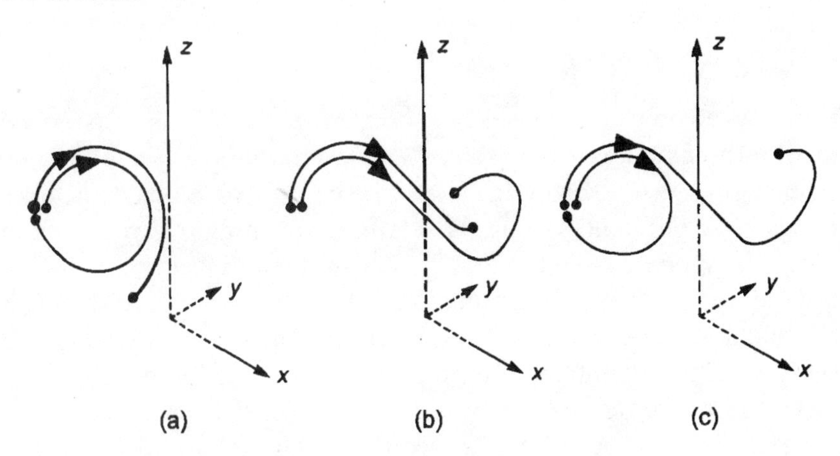

FIGURE 3.7. POSSIBLE HISTORIES OF TWO NEARBY WEATHER STATES

of the Lorenz dynamics, that gives rise to the different behaviors of the three resulting weather systems.

The foregoing situation shows that while in all three cases the two weather states diverge, thereby implying rather different forecasts of the instantaneous weather, it may still be possible to predict the overall weather regime quite far into the future. It all depends on whether we are in situations (a) and (b) or in situation (c). In the first case, the two initial weather states evolve so as to remain on the same branch of the attractor (the same wing of the butterfly). So they represent the same weather regime. But in case (c), the nearby initial states end up in entirely different regimes. Thus we conclude that although the atmosphere itself is chaotic, it may still be possible to predict overall weather patterns from certain initial states.

A natural impulse at this point is to ask, If we are observing the output of a particular dynamical process, how can we tell solely on the basis of these observations whether or not the system is chaotic? What kinds of tests can we apply to decide whether the system's attractor is really "strange," or if it just looks that way in the short term and will eventually settle down to one of the tamer long-run patterns like a fixed point or periodic orbit? Alternatively, is the system's random-looking behavior due to an underlying deterministic rule like the logistic equation, or are we observing a truly random process being generated by an intrinsically probabilistic mechanism?

STATISTICALLY SPEAKING

When asked how he came to discover relativity theory, Einstein replied that he thought about how things would look if he happened to be riding along on a beam of light. What he was saying is that sometimes things look different if you're inside the system than if you look at what's happening from the outside. This is a good example of what in Chapter One we called "jumping out of the system." And so it is too when it comes to chaotic processes. We have an entirely different set of tests for identifying the presence or absence of chaos if we're given the vector field of the system (the insider's view) than if all we have to go on is the outsider's record of only the system's observed behavior (i.e., its output). To start with, let's take our cue from Einstein and examine the question from the insider's perspective.

- *Period-Three theorem*—To fix ideas, suppose we have a system consisting of a set of entities—insects, people, molecules—whose population level at generation t is represented by the number x_t. Furthermore, let the rule specifying how the population grows as a function of its current level be given by the vector field f. So the population at the next generation is $x_{t+1} = f(x_t)$. With an insider's knowledge of the rule f governing the population growth, we can compute the entire population history as soon as we are given the initial level x_0. In 1973, Tien-Yien Li and James Yorke discovered the following remarkable fact about this kind of system: if there is any sequence of three successive generations in which the population increases for two successive periods and then returns to (or goes below) its original level, the system is chaotic. In more colorful language, this result is usually expressed by the title of the famous Li-Yorke paper, "Period Three Implies Chaos." This result gives us very useful information about one-dimensional systems (i.e., those whose states can be represented by a single real number). Unfortunately, the theorem fails for higher-dimensional systems where more delicate tests are called for.

We've already seen that chaos is an ongoing struggle between stretching and folding. This intuitive picture can be "souped up" to yield a test for chaos in higher dimensions. Here's a baker's-eye view of how to do this.

• *Lyapunov exponent*—Think of a baker at work rolling out a ball of pastry dough. The baker takes some dough, works it into a thin sheet with a rolling pin, folds the dough over itself a time or two and then rolls it out again. To make this process mathematically precise, let's assume that at each stage the thickness of the sheet of dough is halved and its length is doubled, the width remaining the same. And instead of folding the sheet, we'll assume that the sheet is cut in half, with one of the half-sheets (always the same one) placed atop the other. After this cutting and placing, the process then begins again with a sheet the same size as before. The first two stages of this process are shown in Figure 3.8, where the face of a cat is placed on the sheet to make it easier to envision how the rolling, cutting and placing are done.

Now suppose that instead of a cat's face we place two dots, *A* and *B*, on the pastry sheet. The diagram in Figure 3.9 shows two applications of the baker's transformation in this case. Note in the figure how the points *A* and *B* have separated during the course of successive

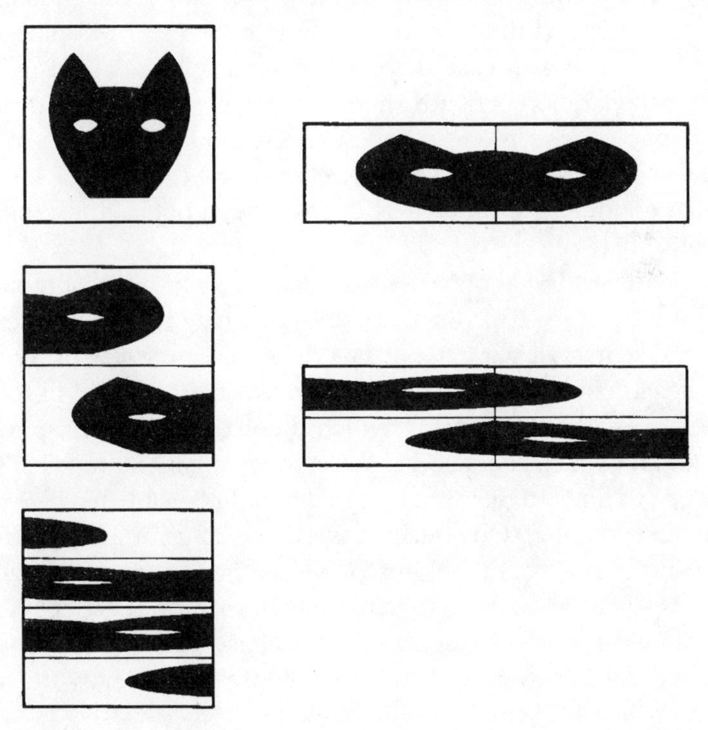

FIGURE 3.8. ROLLING OUT A SHEET OF PASTRY DOUGH

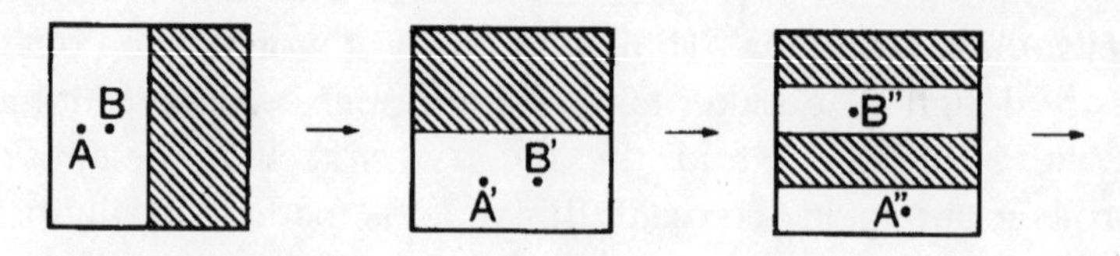

FIGURE 3.9. THE BAKER'S TRANSFORMATION

applications of the transformation. This is a geometric view of what we called earlier *sensitivity to initial conditions*. Two points originally close together get pulled apart by the action of the chaotic baker's transformation. In this pastry situation, there are two directions in which the points can move: up/down and left/right. So there are two rates of separation, one in the vertical direction, the other in the horizontal. These rates are characterized by two numbers, called the *Lyapunov exponents* for the process. And if either of these numbers is positive, indicating a positive rate of separation of the initial points in a given direction, we say the system is chaotic. The situation can easily be extended to the case of a system having an n-dimensional state space, in which case there are n such numbers measuring the degree of separation of nearby initial points in each of the n directions in the space. Since there is one such Lyapunov exponent for each degree of freedom of the system, we will have the kind of sensitivity to initial conditions indicative of chaos if even one of these numbers is greater than zero.

It's clearly asking a lot to demand that the entire rule of motion for the system be given. In fact, for the vast majority of physical and human systems, all we have at our disposal is a set of observations (i.e., measurements) about how the system has behaved in the past. In such cases the very goal to which theoretical science aspires is to produce an explicit characterization of the rule generating what's been seen (i.e., to find a dynamical rule that could plausibly have given rise to the observations). Obviously, it would be of great interest to know if the kind of rule we're looking for should have a strange attractor or not. This situation leads us to take the external, or outsider's, view of the system, in which we seek tests for chaos based solely upon the measured data. Here is one such test involving only how the system behaves when it's actually on the attractor.

- *Correlation dimension*—In a chaotic process, the trajectory from almost every starting point wanders all over the state space. But not all parts of the state space are created equal; some regions are visited more frequently than others. This fact provides the basis for a chaos test based solely on a system's observed behavior.

Consider for a moment the classical attractors, a fixed point and a limit cycle. Regardless of the dimension of the overall state space, the fixed point has dimension 0, while the limit cycle, being a simple closed curve, has dimension 1. These numbers are the *geometric dimension* of the attractor. At the other end of the scale is the situation in which the system is truly random. In this case, every point of the state space is eventually visited, leading to the attractor's having the same geometric dimension as the state space itself. Chaotic systems with their strange attractors lie somewhere in between. For such systems, the attractor is clearly not such a primitive geometrical object as a point or a simple curve. Yet it's still a proper subset of the overall set of states.

Suppose we look at a trajectory that's been moving on its attractor for some time. Let's sample a set of points from this trajectory and compute the statistical correlation between these sample values. This calculation results in a number called the *correlation dimension* of the system, a number that measures the degree to which the system's attractor "fills up" the space of states. Note carefully that this quantity does not usually equal the geometric dimension of the attractor. This is because the correlation dimension weights the points on the attractor according to how frequently they are visited. The geometric dimension, on the other hand, attaches the same weight to each point of the attractor regardless of how often it's visited. A telltale sign of chaos is when the correlation dimension turns out to have a noninteger value much greater than 1.

To illustrate the general idea, suppose we take the data from our logistic map and plot it in the following way. First take the values obtained at times $t = 0$ and $t = 1$. Call these values x_0 and x_1, and think of them as the coordinates of a single point in the two-dimensional x–y plane. So the x coordinate of this point is the number x_0, while the point's y coordinate is x_1. Now do the same thing for the pair of values (x_1, x_2), which are the observed outputs from the logistic rule at times $t = 1$ and $t = 2$, respectively. Continuing in this fashion, we obtain a scattering of points in the x–y plane. Figure 3.10 shows the

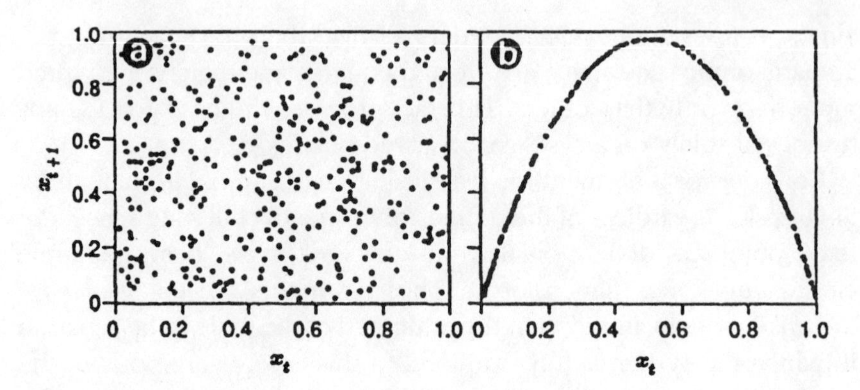

FIGURE 3.10. (A) A RANDOM SCATTER; (B) THE LOGISTIC PARABOLA

possibilities. If the original data points came from a truly random process, then we would expect to see the picture shown in part (a). In this case the points are scattered more or less uniformly throughout the plane. But our data did not come from a purely random process. Rather, it was generated by a very simple, very deterministic rule. So what we actually see is the picture in part (b). This distribution of points is not random, as it certainly does not fill up the two-dimensional plane. This fact suggests the presence of a strange attractor. The correlation dimension sketched earlier is the generalization of this simple idea to settings in which the system's output lives in an n-dimensional world rather than in the one-dimensional universe of the logistic rule.

So with this as our final test for chaos, let's leave the theory behind for a while and turn our attention to some real-life situations in which we can put these rather abstract notions to productive use.

BULLS, BEARS AND BEER

In a little-read and even less well-understood 1900 doctoral dissertation addressing price fluctuations on the Paris bond market, French mathematician Louis Bachelier introduced what has subsequently come to be called the *efficient markets hypothesis* (EMH). Roughly speaking, the EMH states that the driving force behind price changes for any commodity is new information coming into the market from the outside world, and that traders process this information so efficiently that prices adjust instantaneously to this news. So, since the news itself is assumed to appear randomly, prices move in a random fashion as well. This in turn implies that no trading scheme based upon publicly available information can consistently outperform the overall market averages, since any news affecting a firm is instantly discounted into the firm's stock price. Put another way, no amount of processing of past data involving prices, volumes of trade or any other statistic can serve as the basis for a trading strategy that will consistently turn a better rate of profit than what you'd get if you simply held a market portfolio consisting of all the stocks making up a market index like the Dow Jones Industrial Average.

If the EMH is indeed true, then there can be no exploitable structure in stocks and commodities price histories; price movements are truly random. Recently, William Brock, a mathematical statistician from the University of Wisconsin, has jumped all over this hypothesis. Along with others like Blake LeBaron, José Scheinkman, Doyne Farmer and Edgar Peters, Brock has examined long-term price histories of stocks and commodities using some of the tests outlined in the last section, looking for signs that prices are being generated by chaotic mechanisms. This work has focused on two central questions: (1) Do stock price fluctuations come about as the result of an underlying chaotic rule of behavior? (2) If so, can we make use of this structure in the data to develop successful trading strategies? Here it's important to recall the difference between a chaotic process and true randomness. Chaos involves a deterministic mechanism that generates the *appearance* of randomness; a genuine random process has no such deterministic underpinning. So if we can convince ourselves that stock prices are chaotic, then there is a deterministic needle in the haystack to look for; otherwise there isn't. Let's take a few pages to see what kinds of answers Brock & Company have come up with.

Figure 3.11 shows the price movements of the Standard & Poor's Index of 500 stocks (the S&P 500) from January 1950 through July 1989, adjusted for inflation. The S&P 500 index shows a wavelike behavior and seems to be characterized by periods when prices stay consistently high or low. Looking at similar measures of the British, German and Japanese markets, we find analogous overall patterns. With the exception of a period in the 1970s when the British economy

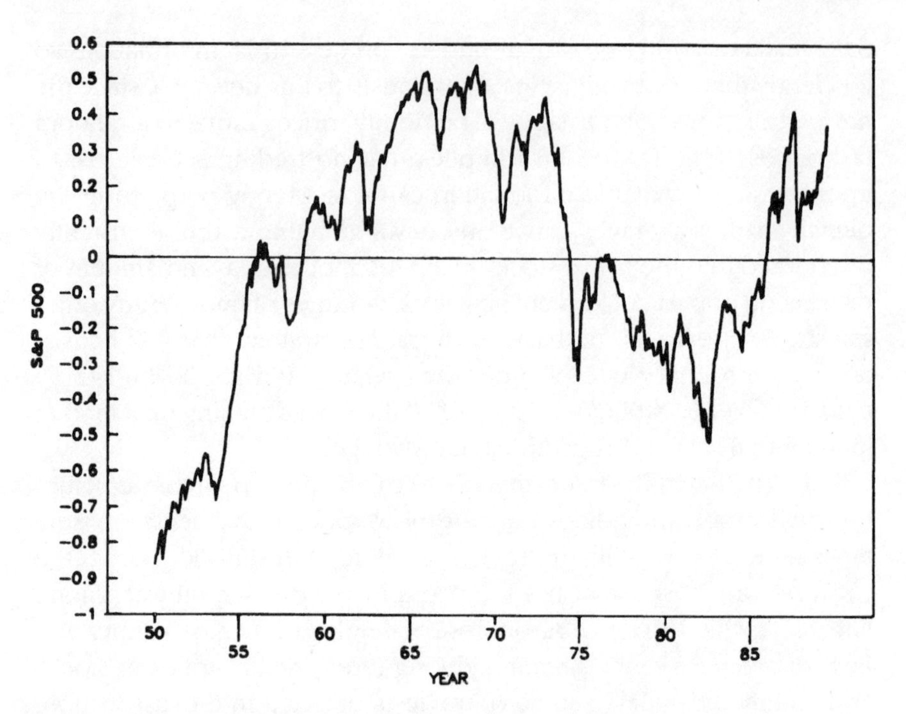

FIGURE 3.11. THE S&P 500 INDEX, JANUARY 1950–JULY 1989, ADJUSTED FOR INFLATION

was falling apart and the pound was under intense pressure, British stock prices tend to follow the economy's internal growth rate. The German market, on the other hand, shows remarkable stability over the entire period. But in the Japanese market there are periods of stable growth followed by accelerated stock values and a collapse back to steady growth. This same Japanese pattern repeated itself in the late 1950s, the late 1960s and again in the second half of the 1980s. The first question then is to determine whether any of these price movements can be convincingly laid at the doorstep of chaos. This is where our tests for chaos come into play.

The initial line of attack on the chaos question involves computing the correlation dimension of each of the price histories. For this, we assume that the one-dimensional series of prices is merely a slice of a higher-dimensional surface, the system's overall attractor. We next postulate the dimension of the smallest space within which this attractor can be embedded. To illustrate this notion of embedding, think of a sphere. It cannot be embedded into a plane, but requires three dimensions. Analogously, the number of dimensions needed to completely encompass the attractor is called the *embedding dimension* for the process. After picking (guessing) an embedding dimension R, we then use the procedures discussed earlier to calculate the statistical correlation of successive data points in this embedding space. As we increase R, we obtain different estimates $C(R)$ for the correlation dimension. If these estimates eventually settle down to a fixed value, we take that limiting value to be the true correlation dimension of the data. Figure 3.12 shows the results of such a calculation for the S&P 500 data.

As we see, the correlation dimension here looks to be about 2.33, a noninteger number greater than 1—strongly indicative of underlying chaos lurking in the data. Table 3.1 gives the outcome of this same kind of calculation for each of the stock indexes. These results suggest that there may well be a chaotic mechanism underlying the price-setting process in each market.

We can check these conclusions by calculating the Lyapunov exponent for each of the price histories. Using monthly returns on the four markets, calculation of the largest Lyapunov exponent yielded the results displayed in Table 3.2.

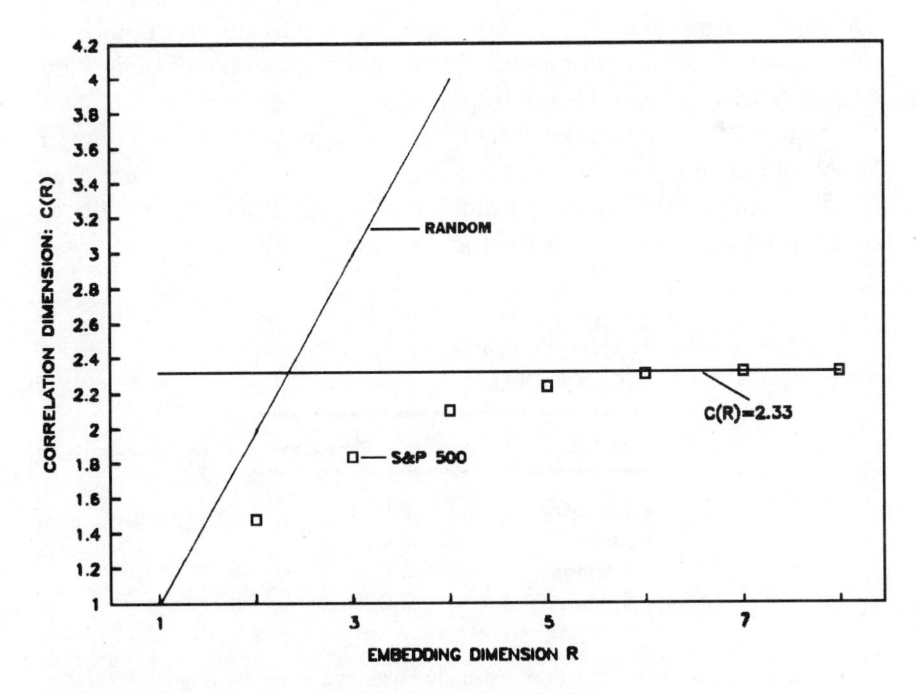

FIGURE 3.12. CORRELATION DIMENSION FOR THE S&P 500 INDEX

TABLE 3.1.
CORRELATION DIMENSION OF STOCK INDEXES

Index	*Correlation Dimension*
S&P 500	2.33
Japan	3.05
Germany	2.41
Britain	2.94

These exponents can be interpreted in two different ways. On the one hand, the Lyapunov exponent for the U.S. market says that we lose predictive power at the rate of 0.024 bits* per month. So if we can measure the initial state to one bit of precision, or about a third of a decimal digit, we will lose all predictive power after 42 months (= 1/0.024). The analogous "forgetting factor" for the other markets is 36 months for the U.K., 44 months for Japan and 60 months for Germany. But there is also a "backward-looking" interpretation of these numbers. Since the U.S. market loses all memory of the initial state after 42 months, this means that market activities 42 months or more apart are no longer related; the market has completely forgotten all past activity after a 42-month period.

The foregoing results certainly seem to implicate chaos as a major factor influencing price fluctuations on the world's stock markets. This leads to our second main question: can we use this information to develop winning ways for our stock market plays?

TABLE 3.2.
LYAPUNOV EXPONENTS FOR STOCK INDEXES

Index	*Lyapunov Exponent*
S&P 500	0.024
Japan	0.022
Germany	0.016
Britain	0.028

* For technical reasons, information is usually measured in binary digits, or "bits." One decimal digit is approximately three bits.

In an extensive series of statistical experiments on stock price data, Brock and his collaborators have concluded that although stock returns do appear to exhibit chaotic structure, the departure from pure randomness is probably too small to be usefully employed in any predictive scheme aimed at beating the market. So it looks as if the good news is that stock price fluctuations do reflect an underlying deterministic, yet chaotic, rule. The bad news, however, is that the departure from true randomness in the output of this rule appears to be too little to serve as the basis for a successful trading scheme. Let me note, however, that the latter conclusion may be reversed if we look at other commodities. For instance, there does appear to be exploitable structure in foreign currency prices. Details of these and other studies can be found in the references cited in the To Dig Deeper section for this chapter.

Stock prices are ultimately determined by the collective effect of many individual human decisions. So as a second foray into the jungle of real-world chaos, let's consider a different kind of decision-making environment—the management of a business enterprise.

In order to reach a widespread market, it's customary for industries to employ a hierarchical distribution system with dealers at many different levels. For example, the basic hierarchy of a demand-supply system for beer distribution consists of the following levels:

- A *distributor*, who receives the beer from the factory and ships it to the main markets
- Regional *wholesalers*, who receive the beer from the distributor and allocate it to local outlets like supermarkets, liquor stores and bars
- *Retailers*, who disperse the products to the consumers

Taken together, we'll call this collection of beer suppliers the *dealers*.

To guard against unpredictable fluctuations in demand and supply, each dealer maintains an inventory. Besides ensuring the availability of beer to the end-consumer, the hierarchical distribution system is meant to facilitate swift restocking if a dealer's inventory runs low. The chain should also function as a buffer to protect the production line from fluctuations in consumer demand. Thus, seasonal and other low-frequency components of the variation in demand should propagate back toward the brewery in a damped fashion.

Figure 3.13 shows the basic structure of the simplified beer distribution system, or *supply line*, that we'll consider here. Orders for beer propagate from right to left, while products are shipped from stage to stage in the opposite direction. Since the processing of orders and the production and shipment of beer involves time delays, we'll assume there is a communication delay of one week (one time-period) from one stage to the next. In the same way, we assume it takes one week to ship beer between two adjacent sectors. Finally, we take the production capacity of the brewery to be unlimited, with a production time of three weeks from the time an order is received from the distributor.

To keep things as simple as possible, all the results reported here assume a customer demand of four cases of beer per week until week five, at which time demand increases to eight cases per week. It is then maintained at this level for the rest of the process. At the beginning of the process, there are assumed to be twelve cases of beer in each dealer's inventory.

The way the process works is as follows: Each week customers order beer from the retailer, who ships the requested quantity out of inventory. The retailer adjusts the order that he places with the wholesaler in response to variation in customer demand and to other pressures. As long as he has sufficient inventory, the wholesaler ships the beer that's been requested. Orders that cannot be met are kept in backlog until delivery can be made. Similarly, the wholesaler orders and receives beer from the distributor, who in turn orders and receives from the brewery. And to keep things as simple as possible, we'll assume that orders cannot be canceled nor can deliveries be returned.

Because of the costs of holding inventory, stock levels should be

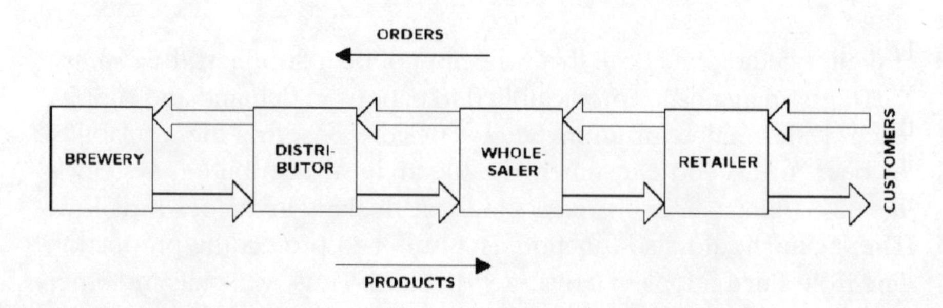

FIGURE 3.13. THE PRODUCTION-DISTRIBUTION SYSTEM IN THE BEER GAME

kept as low as possible. On the other hand, failure to deliver immediately may force customers to look to alternative suppliers. For this reason, penalty costs are assessed for accumulating a backlog of unfilled orders. Therefore, each stock manager must attempt to keep his inventory at the lowest possible level while at the same time avoiding a "stockout." If the inventory begins to fall below the desired level, extra beer must be ordered to rebuild the inventory. If stocks begin to accumulate due to a falloff in demand, the order rate must be reduced. In the experiments discussed here, inventory holding costs are taken to be $0.50 per case per week, while the cost of a backlog is set at $2 per case per week.

The production-distribution system we have just described constitutes the rules for what has come to be called the Beer Game in many graduate business schools in the United States and Europe. In these schools, MBA students play the Beer Game by taking on the role of one or another of the four dealers in the overall distribution chain. The players then try to minimize their cumulative costs over the duration of the game (usually forty weeks).

The decision made by each player each week is the amount of beer to be ordered from that dealer's immediate supplier. The dealers can base their ordering decisions on all information locally available to them (e.g., the current level of their inventory/backlog, previous values of these quantities, expected orders, and anticipated deliveries). In addition, the dealers can use their overall conception—their mental model—of the way the distribution chain works.

Figure 3.14 displays a typical outcome of the Beer Game, showing the variation in effective inventory (i.e., the actual inventory minus the backlog) for the different sectors. Note that a negative effective inventory represents a backlog of unfilled orders.

Generally speaking, a play of the Beer Game is characterized by large-scale oscillations that grow in amplitude from the retailer to the wholesaler and from the wholesaler to the distributor. So by the time the original stepwise increase in customer orders reaches the brewery, it typically leads to an expansion of production by a factor of six or more. Another feature of these games is the increase in orders, which propagates in a wavelike fashion down the chain, depleting the inventories one by one until it's finally reflected at the brewery, at which point the large surplus of orders placed during the out-of-stock period is produced. These features clearly suggest a strong amplifica-

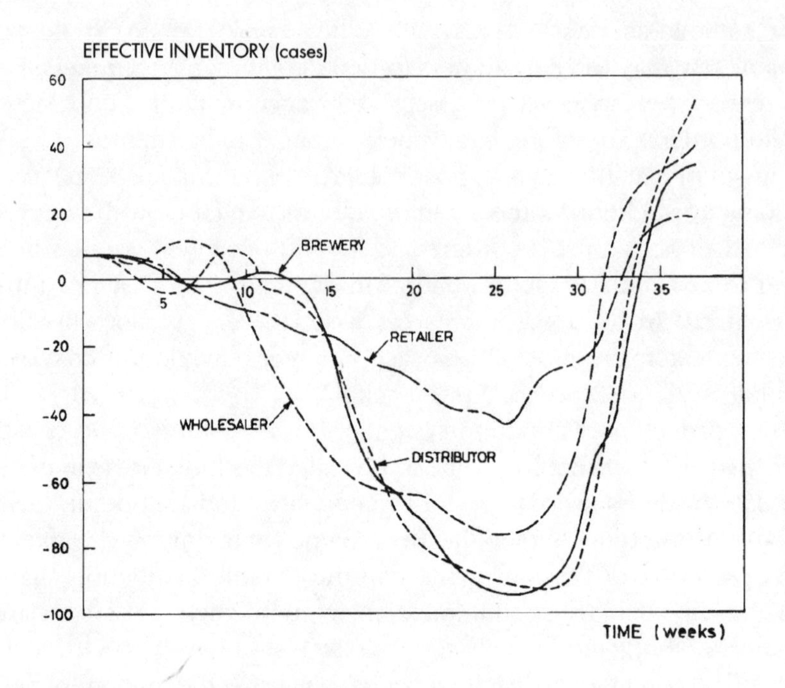

FIGURE 3.14. A TYPICAL OUTCOME OF THE BEER GAME

tion mechanism at work in the system. At the same time, the behavior is restricted by various disproportionalities (i.e., nonlinearities). For instance, we have a pronounced nonlinear relationship between orders and shipments. Together with the relatively high number of state variables, these nonlinearities can generate an extraordinary variety of complex dynamical behaviors.

The amplification process seen in the beer distribution chain is connected with the built-in time delays involved in communicating from one sector to the next. Assume that a particular sector suddenly experiences a significant increase in demand. To discover whether the change in demand is of a more permanent character, players usually hesitate a little before adjusting their own orders by a similar amount, since the very purpose of maintaining an inventory is to absorb rapid changes in demand. However, because of this hesitation the built-in communication and shipping delays ensure that the demand then exceeds inventory replacements for several weeks. So during this period the inventory level goes down. As a result, the players must increase their orders beyond the level of the immediate incoming orders to build the inventory back up to its desired level. As the players come to realize that the increase in demand is of a more enduring nature, they generally increase their orders even more with an eye toward rebuilding their inventory.

The amplification phenomenon in the beer system is a direct consequence of the structure of the distribution chain. But it's important to realize that the beer-distribution chain can be operated in a stable manner. In fact, experience shows that many players are capable of doing just that, and only about a quarter of the participants use ordering policies leading to chaos. Large-scale oscillations are always observed in the transient behavior leading to the system's attractor, however. And in all cases the ordering decisions turn out to be far from optimal, incurring costs exceeding the theoretical minimum by more than a factor of four! Now let's look at some results arising out of a dynamical model created to formalize much of the empirically observed results obtained in more than three decades of play.

First of all, we define a stock adjustment parameter α. This quantity represents the fraction of the inventory shortfall that the participants order in each round. So, for example, if the retail demand is for 100 cases and only 50 cases are available in the inventory, the shortfall is 50 cases. A stock adjustment value of $\alpha = 0.4$ would then result in an order of 20 cases (0.4×50). Initially you might wonder why a dealer wouldn't always order what's needed to cover the current shortfall, a policy that corresponds to setting $\alpha = 1$. The reason is that there's a penalty to be paid for overshooting the anticipated future demands, thus building up too large an inventory.

Similarly, we define the quantity β to be the fraction of the supply line, starting from the brewery, that dealers take into account when placing their orders. So, for example, if $\beta = 1$, the players fully recognize the supply line and do not double-order, while if $\beta = 0$, orders placed are forgotten until the beer arrives. Now for the promised results.

The Beer Game dynamical system contains twenty-seven state variables, quantities such as expected demands, inventory levels and order rates for each of the dealers. Compared with real managerial systems, the model is a vast simplification; but compared with most physical systems investigated in the world of nonlinear dynamics, the model is very complicated. In certain regions of values for α and β, the distribution system has three positive Lyapunov exponents. Therefore we might expect the system to display an unusually complicated spectrum of behaviors. Figure 3.15 shows the distribution of behavior types in the α-β plane. Here the results are plotted using a gray-scale code: light gray indicates stable behavior, dark gray represents aperiodic behavior and black denotes periodic behavior.

A closer inspection of Figure 3.15 shows several regions of unstable behavior separated by "fjords" of stable behavior. For instance, in the regions around $\beta = 0.50$ and $\beta = 0.70$ the model is stable for all values of α, while the narrow peninsula near $\beta = 0.72$ contains only small-amplitude periodic solutions. The other regions of unstable behavior are dominated by large-amplitude fluctuations. The occurrence of unstable behavior is most clearly seen in the lower right corner, where α is large and β relatively small. Therefore, to stabilize the distribution chain it's necessary to use an ordering policy in which inventory discrepancies are adjusted relatively slowly and a significant fraction of the supply line is taken into consideration. However, β should not be too large, since a large value of β increases the costs. This is because the system will then stabilize in a state for which the inventories are negative.

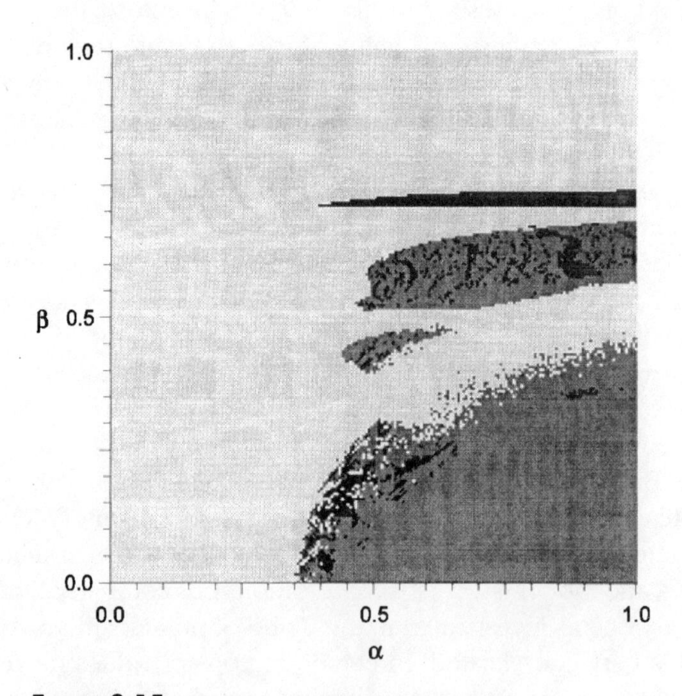

FIGURE 3.15. BEHAVIORAL MODES IN THE α–β PLANE

The foregoing account has only scratched the surface of the many fascinating lessons to be learned about complex dynamics from the Beer Game. We encourage readers to consult the work cited in the To Dig Deeper section for the details. Now let's try to summarize what we've discovered so far about the uncommon sense of chaos and randomness.

COMPUTING THE COSMOS

Earlier we noted that the 18th-century French mathematician and physicist Pierre-Simon de Laplace, caught in the grip of the Newtonian view of the universe as one big clock, made the astounding claim that given at one instant the positions and velocities of all the particles in the universe, a superior intellect could predict with complete certainty what every particle would be doing forever after. Shades of Archimedes and his lever that could move the world! Leaving aside the practical difficulties of carrying out such a calculation, Laplace's bold assertion marked the highwater mark of the idea that the universe is a strictly deterministic machine governed by the eternal laws of mechanics as bequeathed to us by Newton.

Senseless as it seems today, less than two centuries after its proclamation, Laplace's mechanistic view of the world profoundly influenced the attitudes of most scientists of his day. It stood in bold contrast to the dogma Newton replaced, Aristotle's picture of the cosmos as a gigantic organism, a view that had stood unchallenged for over two millennia. And the sharpest point of contrast lay in the concept of free will. A machine can have no free will; its future is rigidly determined from its starting point as it chugs along in accordance with the dictates of Newton's universal recipe. In this rather bleak view of things, the entire history of the universe is written in a vast cosmic book, and the idea of an omnipotent, all-knowing God is reduced to that of a universal librarian merely turning the pages of this gargantuan tome. But along came chaos.

And what does the existence of chaotic processes have to tell us about Laplace's picture of the universe as a machine? The extremely limited predictability of chaotic systems, coupled with their appearance in every nook and cranny of day-to-day life, strongly suggests that the universe is incapable of predicting the future behavior of even a small part of itself. This means that if we want to know what's going to happen with the weather or the stock market over an appreciable length of time, about the best we can do is to turn the system on and just watch it unfold. These systems seem to be their own fastest simulator.

So even if we're willing to accept Laplace's strictly deterministic account of natural and human affairs, the future states of the universe are "open," in the sense that they cannot be predicted before they actually occur. Many have latched on to this fact as an argument for the reality of human free will. Others have suggested that it introduces an element of creativity into nature, endowing the cosmos with an ability to generate something that's not already implicit in its earlier states. Regardless of what interpretation we favor, the one thing that can be said for certain is that the existence of such intrinsically unpredictable processes means that we will never be able to prove that the future of the universe is fixed. If there is a cosmic book containing the history of each and every particle and person, the last chapter of that magnum opus will remain forever unwritten.

Part of the message of chaos is that complicated behavior need not require complicated dynamics. Even the simplest sort of nonlinear behavior, like that arising from the Circle-10 or the logistic rules, can lead to behavior defying any attempt to compactly summarize and express the overall pattern. A central role in arriving at these results was played by the fact that it's impossible, practically speaking, to know a system's *exact* initial state; it's effectively uncomputable. But when it comes to surprises, it's not just in the realm of chaos that uncomputability rears its ugly head.

Our focus in the last two chapters has been on dynamical systems, mathematical style. But manifolds and vector fields are not the only way to represent dynamical processes. Instead of representing them as symbols on a piece of paper or a blackboard, we can also think of them as strings of binary digits inside a computing machine. This point of view dramatically changes how we regard such processes, especially the kinds of questions we can ask about them and the sorts of responses we're ready to dignify with the appellation *answer*. So for the next chapter or two we leave the cozy confines of traditional mathematics, shifting our venue to that handmaiden of modern science and the modern scientist, the digital computer.

FOUR

THE LAWLESS

Intuition: All real-world truths are the logical outcome of following a set of rules.

"Damn it all, we want to get at the truth" [said Lord Peter Wimsey.]
"Do you?" said Sir Impey drily. "I don't. I don't care twopence about the truth. I want a case."
—DOROTHY L. SAYERS

God offers to every mind its choice between truth and repose. Take which you please—you can never have both.
—RALPH WALDO EMERSON

Prove all things; hold fast that which is good.
—NEW TESTAMENT, *I THESSALONIANS*, V, 21

THE POWER OF PARADOX

Exhibit A

Article V of the U.S. Constitution gives conditions under which the Constitution itself may be amended. Specifically, it states that whenever at least two-thirds of both the House and the Senate agree, a constitutional amendment may be enacted, subject to ratification by three-fourths of the state legislatures. Now consider whether Article V may authorize its own amendment or repeal. Can a rule that allows the changing of other rules also admit its own change? Note that this is an "endo" paradox, in that it pertains to changing the rules of the system by using a rule that itself is part of the system.

We have here a legal version of the so-called paradox of omnipotence. Consider the situation where the Congress has the power to make any law at any time. If the Congress indeed has such powers, then can the legislators limit their own power to make law? If they can, then they can't—and conversely. So we may say that either there is a law that the Congress cannot make or a law that it cannot repeal.

Clearly, if we allow the assumption of an omnipotent Congress to stand, then we come to a genuine paradox. But such a postulate implies the affirmation and the negation of the idea that Congress can limit its own power irrevocably. This assumption implies a contradiction; thus it is false. And since its being false does not imply its truth, we can call the postulate of an omnipotent Congress false with finality.

By the view that such an omnipotent Congress cannot exist as defined, a Constitutional clause authorizing its own amendment or that actually limits itself by self-amendment is a contradiction. So we conclude that amendment clauses are immutable except by illegal or extra-legal means like a revolution. In other words, they can only be amended if we take an "exo-legal" view of the situation, in effect allowing the system to examine itself from the outside.

Exhibit B

Early in this century, French mathematician Philip Jourdain distributed calling cards printed on both sides with the messages shown in Figure 4.1. Jourdain's calling card paradox is a variant of a paradox noted by 14th-century Venetian philosopher Jean Buridan. According to Buridan, Socrates makes the single statement "What Plato says is false." Similarly, Plato makes the statement "What Socrates says must be true." The combination of these two statements immediately leads to the conclusion that what Socrates says must be both true and false at the same time. Both Jourdain's and Buridan's double-edged paradoxes are illustrations of the famous Liar, or Epimenides, paradox

FRONT

The statement on the other side of this card is true.

BACK

The statement on the other side of this card is false.

FIGURE 4.1. PHILIP JOURDAIN'S CALLING CARD

reported by Saint Paul in his epistle to Titus, which involves the Cretan Epimenides making the statement "All Cretans are liars."

What's interesting about the these Liar paradoxes is that the individual statements taken separately make perfectly good sense. But like mixing nitric acid and glycerin, things blow up in your face when the two statements are combined.

Exhibit C

The Dutch graphics artist M. C. Escher was famous for the creation of engravings that are like formally constructed logical paradoxes. Among the best known of these visual paradoxes is the 1948 lithograph *Drawing Hands*, shown in Figure 4.2. It shows a pair of hands, each of which is drawing the other. Moreover, both hands are depicted on a piece of paper, which is itself tacked onto a drawing board.

This lithograph contains several paradoxical aspects. First of all, there is the self-referential circularity created by each hand's drawing of the other. But there is also the artistic contradiction between the two-dimensionality of figurative drawing and the three-dimensional world that the drawing purports to represent. Thus, in *Drawing Hands*, as in many of his other works, Escher is stating that all drawing is a form of illusion. And in each of his drawings, Escher illustrates this deception with a visual logic so forceful that the viewer cannot possibly escape the contradictory effects the chain of logic engenders.

Exhibit D

In 1971, Bela Julesz, a researcher at the AT&T Bell Laboratories, reported the remarkable discovery that three-dimensional images

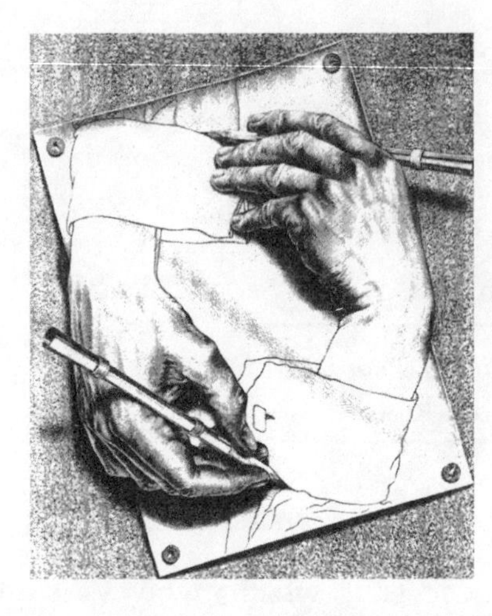

FIGURE 4.2.
DRAWING HANDS BY M.C. ESCHER

could be generated by scattering black dots on a flat piece of paper. An example is shown in Figure 4.3. At first glance, this figure looks about as boring as a figure can be, essentially just a random scattering of dots. But the surprise comes when we see a real three-dimensional

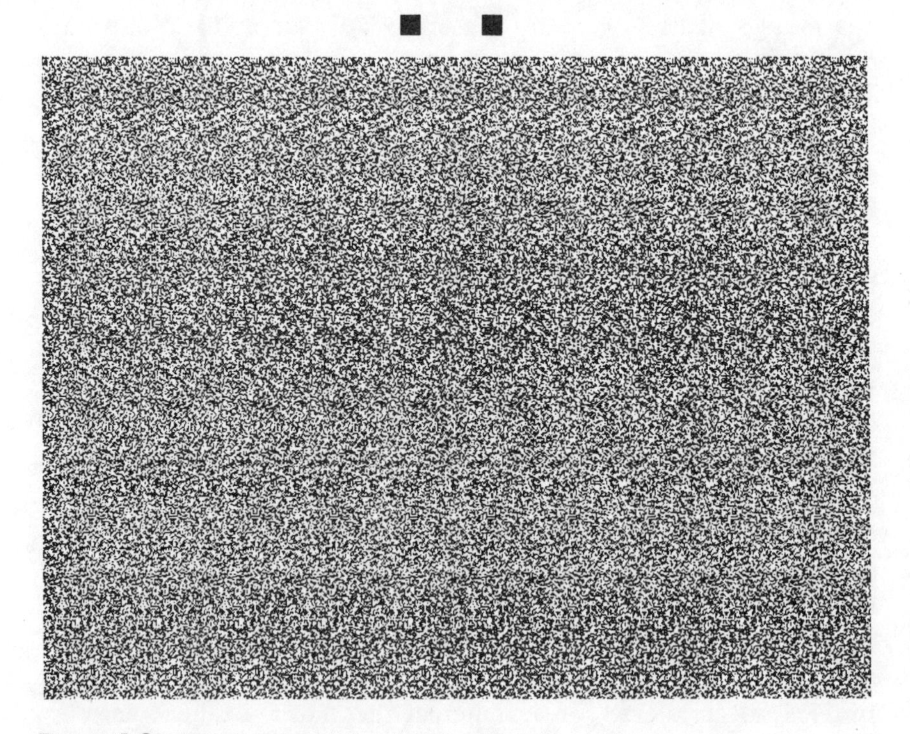

FIGURE 4.3. A JULESZ FIGURE

structure jumping out at us from beneath this seemingly patternless façade. Here's how to see this three-dimensional image.

First note the two black boxes at the top of the image. Stare at these boxes for a while, letting your eyes diverge until you see three boxes. The one in the middle is just like the phantom "sausage" you see when you point your two index fingers at each other and defocus your eyes. Once you can see the third box in the middle, shift your gaze downward *(without refocusing your eyes)* and stare at the dots. Shortly, you should see an image form in the center of the figure that looks like the top halves of six golf balls arranged in a circle so that their sides are touching. The balls seem to be resting on a set of circular ledges whose common center is the center of the circle made by the balls. (Note: It may take a bit of time for this image to form, so if you don't see it right away, keep trying. Like a good—but subtle—joke, when you finally get it the reward will make the effort all worthwhile.)

When we look at an object in the real world, the retina of each eye registers a slightly different image. The brain then fuses these two images to create the perception of a solid object. You may recall the 3-D movies of the 1950s, in which specially colored red and blue glasses were used to project different images from the screen onto each retina. With that technique, the eye looking through the red lens sees only the blue part of the red-and-blue screen image, while the eye looking through the blue lens sees only the red part of the image. What these Julesz figures illustrate is that sometimes it's possible to obtain the same stereoscopic effects without having to use 3-D glasses.

So what are Exhibits A through D trying to tell us about paradox, surprise and truth? Contemplation of these lofty matters leads to the conclusion that by following a set of rules, we don't always get to the truth. In some cases, as with the problem of an omnipotent Congress, the rules can lead to contradiction. In other cases, truth seems to escape being fenced in by a set of logical rules, as was seen verbally in the Liar paradox of Exhibit B and visually in the Escher graphic of Exhibit C. Finally, the Julesz figure of Exhibit D shows that the same pattern can contain more than one truth. But by following a set of rules (in this case, the rules employed by the human visual-processing system), we are able to see only one of these truths at a time—literally.

The common element here is the idea of getting at the scheme of things by following a set of rules, a formula, a recipe. Speaking slightly more formally, this leads us to ask what kind of real-world truths we can hope to uncover by carrying out a computation. When it comes to forming expectations about events of the world, we often employ prescriptions of various sorts that amount to following a set of rules. And it's the end result of these rules that constitute our expectations, which in turn serve as the root cause of surprise. So in this chapter we ask if every real-world truth is, at least in principle, obtainable as the end result of a logical train of reasoning. Or, on the other hand, are there truths that no amount of rule-following will ever enable us to know. This idea leads us to ask if there are limits to the power of rational thought. Are there truths about the world that can never be accessed by carrying out a computation? So our leitmotif in this chapter is the idea of computation as a truth-generating mechanism.

REALITY RULES

Consider the following two sets of arguments:

Argument A

Everybody loves a lover.
George doesn't love himself.
Therefore George doesn't love Martha.

Argument B

Either everyone is a lover or some people are not lovers.
If everyone is a lover, Waldo certainly is a lover.
If everyone isn't, then there is at least one nonlover, call her Myrtle.
Therefore if Myrtle is a lover, everyone is.

Given the well-chronicled human proclivity for fuzzy thinking and logical inconsistency, it probably comes as a bit of a surprise for most of us to discover that both of these chains of reasoning are logically correct. Wouldn't it be nice if we had a logic machine into which we could feed these kinds of statements, turn the handle and have the machine tell us with finality whether the line of argument is logically valid? Such a machine is what logicians call a *decision procedure*. And it was the search for just such a procedure that sparked off Alan Turing's investigation into the foundations of computing. To ease our way into this story, let's focus on the ideas underlying the difference between something's being logically correct—like arguments A and B—and something's being true—like a real-world George who really does not love some real-world Martha.

In logic and mathematics, the road to truth is paved by the stones that nowadays we call a *formal system*. In general terms, a formal system is a collection of abstract symbols, together with a set of rules for assembling the symbols into strings. Such a system has four components:

- *Alphabet*—This is simply the set of abstract symbols itself. These may be as primitive as the symbols o and –. But they could also be much more concrete, such as the characters of the Latin alphabet, along with the symbols for punctuation, logical combination and the like.
- *Grammar*—The grammatical rules of the system specify the valid ways in which we can combine the symbols of the alphabet to form finite strings of symbols (words), as well as how such strings may be combined into larger statements. Statements of this type are termed *well formed*.
- *Axioms*—A set of well-formed statements that are taken as given strings of the system are called the *axioms* of the system. Thus, an axiom is a string of symbols that we accept as a valid statement without its having to be proved.
- *Rules of inference*—These are the procedures by which we combine and change axioms and other well-formed statements into new well-formed strings.

Now suppose we are given a well-formed statement A and ask if it is a logically correct consequence of the axioms of the system. We say that A is *provable* in the system if there is a finite chain of statements

$$M_1 \rightarrow M_2 \rightarrow M_3 \rightarrow \ldots \rightarrow M_n = A$$

where each M_i is either one of the axioms of the system or is obtained from the previous Ms via one of the rules of inference. Well-formed statements for which such a *proof sequence* exists are called the *theorems* of the system. Since the idea of a formal system is so central to our concerns in this chapter, let's look at a concrete example to fix these basic concepts firmly.

Suppose the symbols of our system are the three more or less culturally free objects ★ (star), ✠ (Maltese cross), and ✪ (sunburst). Let the two-element string ✠✪ be the sole axiom of the system, and take the rules of inference to be

Rule I: x✪ → x✪★
Rule II: ✠x → ✠xx
Rule III: ✪✪✪ → ★
Rule IV: ★★ → —

In these rules, x denotes an arbitrary finite string of stars, crosses and sunbursts, while the → (arrow) means "is replaced by." The interpretation of Rule IV is that any time two stars appear they can be dropped from the string. Now let's see how these rules can be used to derive a theorem.

Starting with the single axiom ✠✪, we can deduce that the string ✠★✪ is a theorem by applying the rules of inference in the following order:

→	✠✪	→	✠✪✪	→	✠✪✪✪✪	→	✠★✪
(Axiom)		(Rule II)		(Rule II)		(Rule III)	

Note that when applying Rule III at the final step, we could have replaced the last three ✪s from the preceding string rather than the first three, thereby ending up with the theorem ✠✪★ instead of ✠★✪. You will probably have also noted that all the intermediate strings obtained in moving from the axiom to the theorem begin with a ✠. It's easy to see from the action of the transformation rules for this system that every string will have this property. This is a *metamathematical* property of the system since it's a statement *about* the system rather than one made *in* the system itself. We'll see in a moment that this distinction between what the system can say from the inside (its strings) and what we can say about the system from the outside (properties of the strings) is of the utmost importance for understanding the limitations of formal systems when it comes to getting at real-world truths.

From the foregoing example, if not from the definition of a formal system itself, it's evident that in a very definite sense the theorems of a formal system are already present in the axioms and rules of inference. All a proof sequence does is make explicit that which is already implicit. Thus we can think of a formal system as a compact way of summarizing what may be a very large body of "fact." But these are just formal facts, essentially statements of what kind of logical consequences follow from given assumptions.

But what do Maltese crosses, sunbursts, and stars have to do with anything? Speaking even more specifically, what do these symbols have to do with very definite mathematical things like the sum of the first n positive integers or the angles of a triangle? The answer to this eminently sensible query lies in one word: *interpretation*. Given a particular kind of mathematical structure, we have to make up a dictionary to translate (i.e., interpret) the abstract symbols and rules of the formal system into the objects of that structure. By this dictionary-construction step, we attach a *meaning* to the abstract, purely syntactic structure of the symbols and strings of the formal system. Thereafter, all the theorems of the formal system can be interpreted as true statements about the associated real-world objects. The following diagram illustrates this crucial distinction between the purely syntactic world of formal systems and the semantic world of mathematical objects.

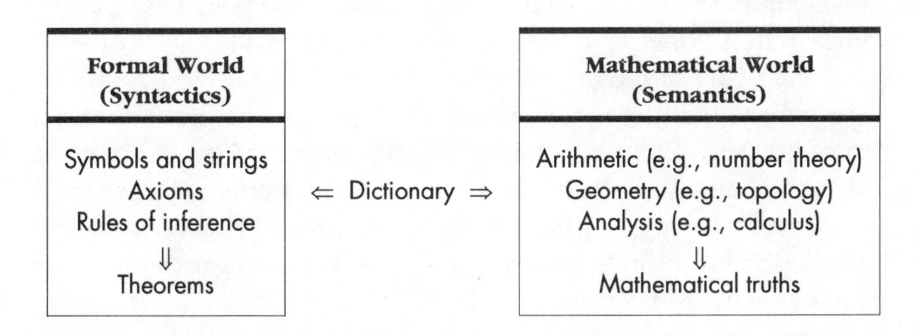

This kind of translation between abstract symbols and real-world things is not just confined to mathematical objects. For example, consider the game of chess. Typically it's played with a collection of chess pieces on a chessboard consisting of black and white squares. We can all agree, I think, that these are indeed real-world objects.

Moreover, the game involves a set of rules specifying how the pieces can move, what constitutes a legal position on the board, how one piece captures another and so forth. This is the real-world version of the game of chess. But there is another version, one existing purely in the world of symbols and syntax (i.e., formal systems), and this version mirrors exactly the real-world game we normally see. Here's what the formal-system version looks like.

A moment's thought shows that the game of chess is really a relationship between one set of abstract symbols (the Black and White pieces) and another set of abstract symbols (the squares of the board). In short, there is nothing crucial about the material embodiment of these symbols insofar as the essentials of the game are concerned. So, for instance, we could assign a set of symbols as in our ★-✠-✡ system to represent the various playing pieces, along with another set of symbols (e.g., the set of positive integers 1, 2, . . . 64) to represent the squares of the board. The grammar of such a system would then specify what strings of symbols (statements) are well formed (i.e., represent valid configurations of pieces on the board; e.g., a Black bishop cannot sit on a white square). Such grammatically correct statements represent the possible states of play at any stage of the game. Moreover, the rules of inference of this formal system are simply the different ways that one well-formed sentence can be transformed into another. In other words, the rules of inference represent the allowable moves at any stage of play. Finally, the sole axiom of the game of chess is the symbol string corresponding to the way the pieces sit on the board at the beginning of play.

So we see that the real world of chess pieces and playing boards can be translated into a formal-world version of the game involving only abstract symbols, rules of inference and axioms. And the same line of argument applies to every other real-world situation that can be described in a finite number of words. We'll return to this point with a vengeance later on.

Let me again emphasize that there are two entirely different worlds being mixed up here: the purely syntactic world of the formal system and the meaningful world of mathematical objects and their properties. And in each of these worlds there is a notion of truth: theorems in the formal system, factually correct statements such as "2 + 5 = 7" or "the sum of the angles of a triangle equals 180 degrees" in the realm of mathematical reality. The connection between the two worlds lies in the interpretation of the elements of the formal system in terms of the objects and operations of the mathematical structure. Once this dictionary has been written and the associated interpretation established, then we can hope that there will be a perfect, one-to-one correspondence between the true facts of the mathematical structure and the theorems of the formal system. Speaking loosely, we seek a formal system in which every truth translates into a theorem, and conversely. Such a system is termed *complete*. We'll consider the degree to which this ideal relationship between the world of symbols and the world of mathematical facts can be approached shortly. For the moment, however, let's stay within the formal world of symbols and rules, looking just a bit deeper into the ins and outs of computing machines and formal rule-based systems.

We are now in a position to describe the problem that stimulated Alan Turing to devise the theoretical gadget now called a *Turing machine*. This question goes under the rubric of Hilbert's Decision Problem. It can be stated as follows: For every formal system *F*, is it possible to find a finitely describable formal system that "decides" any well-formed string in *F*? Loosely speaking, we ask if there is a systematic procedure that will tell us if any given well-formed string of the formal system *F* is or is not a theorem.

For some formal systems, such decision procedures clearly do exist. For example, in the ★-✠-✡ system given earlier, such a decision procedure consists of the following rules: A well-formed string is a theorem if and only if (1) it begins with a ✠, (2) the remainder of the string consists solely of ★s and ✡s and (3) the number of ✡s is not a multiple of 3.

As noted, Turing's attack on the Decision Problem led him to construct the key element in the modern theory of computation, the Turing machine. So let's now turn our attention away from formal logical manipulations and take a longer look at Turing's accomplishment.

MAGIC MACHINES AND BUSY BEAVERS

What is a computation? Oddly enough, despite the fact that humans have been calculating things for thousands of years, a proper scientific answer to this seemingly straightforward query was not forthcoming until 1935. In that year, Alan Turing was a student at Cambridge

University sitting in on a course of lectures in mathematical logic. A central theme of the course was the issue of whether or not there could exist a finite set of rules, in effect a *mechanism*, that would settle the truth or falsity of every possible statement about numbers that could be made in, say, the language of Russell and Whitehead's (in)famous work, *Principia Mathematica*. In short, the question was whether there was a machine into which we could feed any statement about numbers so that after a finite amount of time the machine would spit out the verdict on the statement TRUE or FALSE.

Turing's speculations about what it would mean to have such a mechanical procedure, or *effective process*, for solving this famous Decision Problem led him to develop a mathematical type of computer. This abstract gadget, the Turing machine, provided the first completely satisfactory answer to what it means to carry out a computation.

Turing took the commonsense view of looking at what a human being actually does when carrying out a computation. As it turns out, the distilled essence of computing comes down to the rote following of a set of rules. So, for example, if you want to calculate the square root of 2, you might employ the following rule for creating a set of numbers $\{x_i\}$ that will (hopefully) converge to the quantity $\sqrt{2}$: $x_{n+1} = (x_n/2) + (1/x_n)$. Starting with the initial approximation (i.e., guess) $x_0 = 1$, this rule generates the successively better approximations $x_1 = 3/2 = 1.5$, $x_2 = 17/12 = 1.4166$, $x_3 = 577/408 = 1.4142$. So after just three steps, we have the desired answer correct to four significant figures. For our purposes here, what's important about this so-called Newton-Raphson method for calculating the square root of 2 is not the rapid rate of convergence, but that the procedure represents a purely mechanical, step-by-step process (technically; an *algorithm*) for finding the desired quantity.

The fact that every step in such a procedure is completely and explicitly specified led Turing to believe that it would be possible to construct a machine to carry out the computations. Once the algorithm and the starting point are given to the machine, computation of the sequence of results becomes a purely mechanical matter, involving no judgment calls or interventions by humans along the way. But it would require a special type of machine to accomplish this computational task; not just any mechanical device will do. A large part of

Turing's genius was to show that the very primitive type of abstract computing machine he invented is actually the most general type of computer imaginable. In fact, every real-life computer that's ever been built—or ever will be built—is just a special case that materially embodies the machine that Turing dreamed up. This result is so central to understanding the limitations of machines that it's worth our while to take a few pages to describe it in more detail.

A Turing machine consists of two components: (1) an infinitely long tape ruled off into squares that can each contain one of a finite set of symbols, and (2) a scanning head that can be in one of a finite number of states or configurations at each step of the computational process. The head can read the squares on the tape and write one of the symbols onto each square, replacing whatever symbol happens to be there. The behavior of the Turing machine is controlled by an algorithm, or what we now call a *program*. The program is composed of a finite number of instructions, each of which is selected from the following set of possibilities: change or retain the current state of the head; print a new symbol or keep the old symbol on the current square; move left or right one square; stop. That's it. Just seven simple possibilities. The overall situation is depicted in Figure 4.4 for a Turing machine having twelve internal states labeled A through L. Which of the seven possible actions the head takes at any step of the process is determined by the current state of the head and what it reads on the square it's currently scanning. But rather than continuing to speak in these abstract terms, it's simpler to run through an example in order to get the hang of how such a device operates.

Assume we have a Turing machine with three internal states, *A, B* and *C*, and that the symbols that can be written on the tape are just the two integers 0 and 1. Now suppose we want to use this machine to carry out the addition of two whole numbers. For definiteness, we'll represent the integer *n* by a string of *n* consecutive 1s on the tape. The program shown in Table 4.1 serves to add any two whole numbers using this 3-state Turing machine.

The reader should interpret the table entries in the following way: the first entry is the symbol the head should print, the second element is the direction the head should move, R(ight) or L(eft), while the final

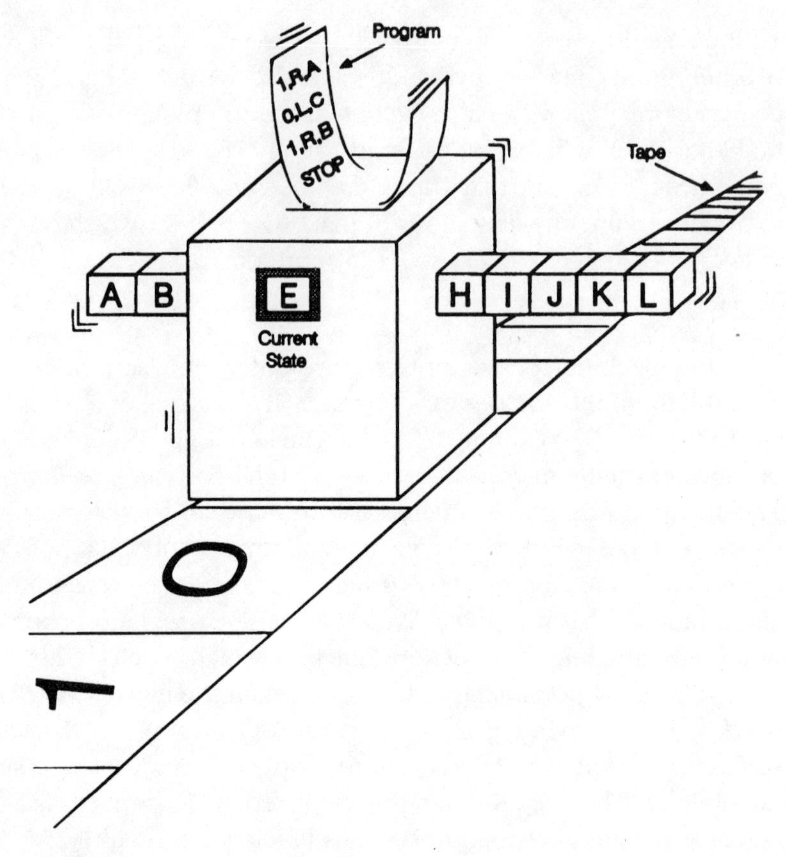

FIGURE 4.4. A 12-STATE TURING MACHINE

TABLE 4.1.
A TURING MACHINE PROGRAM FOR ADDITION

	Symbol Read	
State	**1**	**0**
A	1, R, A	1, R, B
B	1, R, B	0, L, C
C	0, STOP	STOP

element is the state the head should then move into. Note that the machine stops as soon as the head goes into state C. Let's see how it works for the specific case of adding the numbers 2 and 5.

Since our interest is in using the machine to add 2 and 5, we place two 1s and five 1s on an otherwise blank (all 0s) input tape, separating them by a 0 to indicate that they are two distinct numbers. Thus the machine begins by reading the input tape

...	0	0	0	1	1	0	1	1	1	1	1	0	0	0	0	...

By convention, we assume the head starts in state A and reads the first nonzero symbol on the left. Since this symbol is a 1, the program tells the machine to print a 1 on the square and move to the right, retaining its internal state A. The head is still in state A and the current symbol read is again a 1, so the machine repeats the previous step and moves one square farther to the right. Now, for a change, the head reads a 0. The program tells the machine to print a 1, move to the right, and switch to state B. I'll leave it to the reader to complete the remaining steps of the program, verifying that when the machine finally halts, the tape ends up looking just like the input tape above, except with the 0 separating 2 and 5 having been eliminated (i.e., the tape will have seven 1s in a row, as required).

Before looking at the revolutionary implications of Turing's idea, let me pause here to emphasize that Turing machines are definitely not machines in the everyday sense of being material devices. Rather they are "paper computers," completely specified by their programs. Thus, when we use the term *machine* in what follows, the reader should read *program* or *algorithm* (i.e., software) and put all notions of hardware out of sight and out of mind. This abuse of the term *machine* should have been clear from Turing's idea of an *infinite* storage tape, but it's important to make the distinction as clear as possible: Turing machine = program. Period.

Modern computing devices, even home computers like the one I'm using to write this book, look vastly more complicated and powerful in their computational power than a Turing machine with its handful of internal states and very circumscribed repertoire of scanning-head actions. Nevertheless, this turns out not to be the case, and a large measure of credit is due to Turing for recognizing that *any* algorithm

TABLE 4.2.
A CODING SCHEME FOR THE TURING MACHINE LANGUAGE

PROGRAM STATEMENT	CODE
PRINT 0	000
PRINT 1	001
GO RIGHT	010
GO LEFT	011
GO TO STEP *i* IF THE CURRENT SQUARE CONTAINS 0	10100 . . . . . . 01 (*i* repetitions)
GO TO STEP *i* IF THE CURRENT SQUARE CONTAINS 1	11011 . . . . . . 10 (*i* repetitions)
STOP	100

(i.e., program) executable on *any* computing machine—idealized or otherwise—can also be carried out on a particular version of his machine, termed a *universal Turing machine* (or UTM for short). So except for the speed of the computation, which definitely *is* hardware-dependent, there's no computation that my machine (or anyone else's) can do that can't be done with a UTM.

To specify his UTM, Turing realized that not only the input data of the problem, but also the program itself could be coded by a series of 0s and 1s. Consequently, we can regard the program as another kind of input data, writing it onto the tape along with the data it is to process. Table 4.2 shows one of the many ways this coding can be done.

With this key insight at hand, Turing constructed a program that could simulate the action of any other program *P* when given *P* as part of its input (i.e., he created a UTM). The operation of a UTM is simplicity itself.

Suppose we have a particular Turing machine specified by the program *P*. Since a Turing machine is completely determined by its program, all we need do is feed the program *P* into the UTM along with the input data. Thereafter the UTM will simulate the action of *P* on the data; there will be no recognizable difference between running the program *P* on the original machine or having the UTM pretend it *is* the Turing machine *P*.

What's important about the Turing machine from a theoretical point of view is that it represents a formal mathematical object. So with the invention of the Turing machine, for the first time we had a well-defined notion of what it means to compute something. But this then raises the question What exactly can we compute? In particular, is there a suitable Turing machine that will compute every number? Or do there exist numbers that are forever beyond the bounds of computation? Turing himself addressed this problem of computability in his trail-blazing 1936 paper, in which he introduced the Turing machine as a way of answering these fundamental questions.

First of all, let's be clear on what we mean by a number being computable. Put simply, an integer n is said to be *computable* if there is a Turing machine that, starting with a tape containing all 0s, will stop after a finite number of steps with the tape then containing a string of n 1s and all the rest 0s. The case of computing a real number is a bit trickier since most real numbers consist of an infinite number of digits. So we call a real number computable if there is a Turing machine that will successively print out the digits of the number, one after the other. Of course, in this case the machine will generally run on forever. With these definitions in hand, let's look at the limitations on our ability to compute numbers.

It's an easy exercise to show that for a two-symbol Turing machine with n possible states of the reading head, there are exactly $(4n + 4)^{2n}$ distinct programs that can be written. This means that an n-state machine can compute at most this many numbers. Letting n take on the values $n = 1, 2, 3, \ldots$, we conclude that Turing machines can calculate at most a *countable* set of numbers; that is, a set whose elements can be put into a one-to-one correspondence with a subset of the positive integers (the "counting" numbers). But there are uncountably many real numbers; hence, we come to the perhaps surprising result that the vast majority of real numbers are not computable.

This counting argument is one way to show the existence of uncomputable numbers, albeit a somewhat indirect one. Turing himself used a more direct procedure based upon what's known as *Cantor's Diagonal Argument*. It goes like this. Consider the following listing of names Smith, Otway, Arquette, Bethel, Bellman and Imhoff. Now take the first letter of the first name and advance it alphabetically by one position. This gives a T. Then do the same for the second letter of the second name, the third letter of the third name and so on. The result is "Turing." It's clear, I think, that the name Turing could not have been on the original list, since it must differ from each entry on that list by at least one letter.

Turing's argument for the existence of uncomputable numbers follows the same line of reasoning. Suppose you list all computable numbers, written out by their decimal expansions (even though such a list will be infinitely long). Now advance the first digit of the first number, the second digit of the second number and, in general, the kth digit of the kth number. In this way we create a new number. This number cannot have been on the original list, since it differs in at least one position from every number on that list. But by definition, the list contains all computable numbers. Hence the new number must be uncomputable.

From the foregoing arguments, we see that uncomputable numbers are not rara avis in the arithmetic aviary. Quite the contrary, in fact, it's the computable numbers that are the exception rather than the rule. This surprising fact shows that all the numbers we deal with in our everyday personal and professional lives, which by their very nature must be computable, form but a microscopically small subset of the set of all possible numbers. The overwhelming majority of numbers lie in a realm that's impossible to reach by following the rules of any type of computing machine. Now let's look at an amusing example of a specific uncomputable quantity.

The Busy Beaver Game

Suppose you're given an input tape filled entirely with 0s. The challenge is to write a program for an n-state Turing machine such that (1) the program must eventually halt, and (2) the program prints as many 1s as possible on the tape before it stops. Obviously, the number of 1s that can be printed is a function only of n, the number of states available to the machine. Equally clear is the fact that if $n = 1$, the maximum number of 1s that can be printed is only one, a result that follows immediately from the requirement that the program cannot run on forever. If $n = 2$, it can be shown that the maximum number of 1s that can be printed before the machine halts is four. Programs that print a maximal number of 1s before halting are called *n-state Busy Beavers.* Table 4.3 gives the program for a 3-state Busy Beaver, while Figure 4.5 shows how this program can print six 1s on the tape before stopping. (Note: the position of the tape-scanning head is shown in boldface in the figure.) Now for our uncomputable quantity. Define $BB(n)$ to be the number of 1s written by an n-state Busy Beaver program. Thus, the Busy Beaver function $BB(n)$ is the greatest number of 1s that any halting program can write on the tape of an n-state Turing machine. We have already seen that $BB(1) = 1$, $BB(2) = 4$, and $BB(3) = 6$. From these results for small values of n, you might think that the function $BB(n)$ doesn't have any particularly interesting properties as n gets larger. But just as you can't judge a book by its cover (or title), you also can't judge a function from its behavior for just a few values of its argument. In fact, detailed investigation has shown that

$$BB(12) \geq 6 \times 4096^{4096^{4096^{\cdot^{\cdot^{\cdot^{4096^{4}}}}}}}$$

where the number 4096 appears 166 times in the dotted region! So in trying to calculate the value of the Busy Beaver function for a 12-state Turing machine, we quickly arrive at a number so huge that it's effectively infinite. It turns out that for large enough values of n, the value $BB(n)$ exceeds the value of *any* computable function evaluated at that same number n. In other words, the Busy Beaver function $BB(n)$ is uncomputable. So for a concrete example of an effectively uncomputable number, just take a Turing machine with a large number of states n. Then ask for the value of the Busy Beaver function for that value of n. The answer is to all intents and purposes an uncomputable

TABLE 4.3.
A 3-STATE BUSY BEAVER

	Symbol Read	
State	**0**	**1**
A	1, R, B	1, L, C
B	1, L, A	1, R, B
C	1, L, B	1, STOP

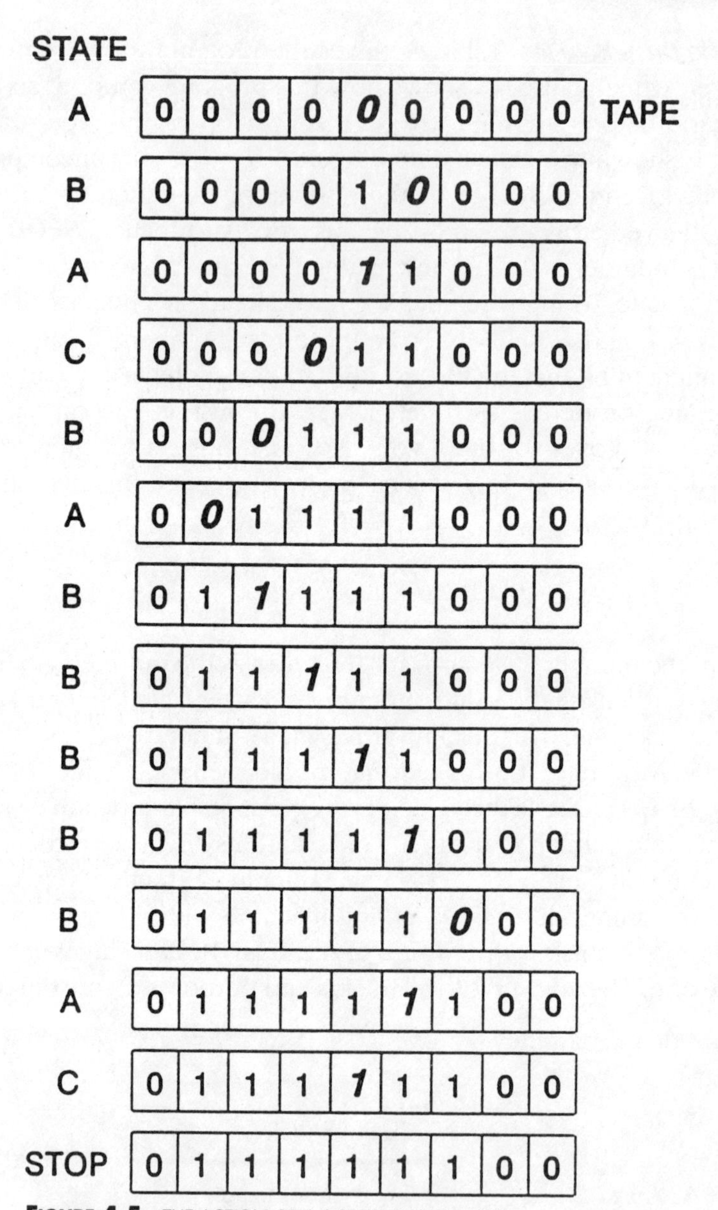

FIGURE 4.5. THE ACTION OF A 3-STATE BUSY BEAVER

number. Now to firmly fix the difference between something's existing and our being able to compute that same something, let's consider another kind of game.

The Turing Machine Game

Assume there are two players, called rather unimaginatively A and B. These players take turns choosing positive integers as follows:

Step 1: Player A chooses a number n.
Step 2: Knowing n, Player B picks a number m.
Step 3: Knowing m, Player A selects a number k.

Player A wins if there is some n-state Turing machine that halts in exactly $m + k$ steps when started on a tape containing all 0s. Otherwise, Player B wins. It's fairly clear that this is a game of finite duration, since once the players have chosen their integers, all we need do is list the $(4n + 4)^{2n}$ Turing machines having n states, and run each of them for exactly $m + k$ steps to determine a winner.

It's a well-known fact from game theory that any game of fixed, finite duration is determined, in the sense that there is a winning strategy for one of the players. In this case, it's Player B. Nevertheless, the Turing Machine Game is nontrivial to play, since neither player has an *algorithm* (i.e., a computable strategy) for winning the game. The proof of this fact relies on showing that *any* winning strategy involves computing a function whose values grow faster than those of the Busy Beaver function. But we already know that $BB(n)$ is uncomputable; hence, this new function must also be uncomputable. The reader is referred to the To Dig Deeper section for further details of this proof.

From our definition of computability, together with the Busy Beaver example above, it's clear that you haven't really computed anything until the computational process terminates—even in the case of real numbers, where any finite computation generally yields only an approximation to the number you're trying to compute. This simple observation leads to a key question in the theory of computation: Is there a general procedure (i.e., an algorithm) that will tell us *in advance* whether or not a particular program will halt after a finite number of steps? In other words, given any Turing machine program P and a set of input data I, is there a single program that accepts P and I and that will tell us whether or not P will halt after a finite number of steps when processing the data I? Note carefully that what we're asking for here is a *single* program that will work in *all* cases. This is the famous *Halting Problem*.

To see that the question is far from trivial, suppose we have a program P that reads a Turing machine tape and stops when it comes to the first 1. So in essence the program says, "Keep reading until you come to a 1, then stop." In this case, input data I consisting entirely of 1s would result in the program stopping after the first step. On the other hand, if the input data were all 0s, then the program would never stop. Of course, in this situation we have a clear-cut procedure for deciding whether or not the program will halt when processing some input tape: the program will stop if and only if the input tape contains even a single 1; otherwise, the program will run on forever. So here's an example of a halting rule that works for any data set processed by this especially primitive program.

Unfortunately, most real computer programs are vastly more complicated than this, and it's far from clear by simple inspection of the program what kinds of quantities will be computed as the program goes about its business. After all, if we knew what the program was going to compute at each step, we wouldn't have to run the program. Moreover, the stopping rule for real programs is almost always an implicit rule of the foregoing sort, saying something like "If such and such a quantity satisfying this or that condition appears, stop; otherwise, keep computing." The essence of the Halting Problem is to ask if there exists any *effective procedure* that can be applied to the program and its input data to tell beforehand whether or not the program's stopping condition will ever be satisfied. In 1936, Turing settled the matter once and for all in the negative: given a program P and an input data set I, there is no way in general to say if P will ever finish processing the input I.

The notion of a Turing machine finally put the idea of a computation on a solid mathematical footing, enabling us to pass from the vague, intuitive idea of an effective process to the precise, mathematically well-defined notion of an algorithm. In fact, Turing's work, along with that of the American logician Alonzo Church, forms the basis for what has come to be called the

Turing-Church Thesis *Every effective process is implementable by running a suitable program on a UTM.*

The key message of the Turing-Church Thesis is the assertion that any quantity that can be computed can be computed by a suitable Turing machine. This claim is called a thesis and not a theorem because it's not really susceptible to proof. Rather, it's more in the nature of a definition, or a proposal, suggesting that we agree to equate our informal idea of carrying out a computation with the formal mathematical idea of a Turing machine.

To bring this point home more forcefully, it's helpful to draw an analogy between a Turing machine and a typewriter. A typewriter is also a primitive device, allowing us to print sequences of symbols on a piece of paper that is potentially infinite in extent. A typewriter also has only a finite number of states that it can be in: upper and lower case letters, red or black ribbon, different symbol balls and so on. Yet despite these limitations, any typewriter can be used to type *The Canterbury Tales, Alice in Wonderland* or any other string of symbols. Of course, it might take a Chaucer or a Lewis Carroll to tell the machine what to do. But it can be done. By way of analogy, it might take a very skilled programmer to tell the Turing machine how to solve difficult computational problems. But, says the Turing-Church Thesis, the basic model—the Turing machine—suffices for every type of problem that is at all solvable by carrying out a computation.

It has probably not escaped the reader's attention that there is a striking parallel between the actions taken by a Turing machine as it goes about performing a computation and the steps one follows in creating a deductive argument leading from premises to conclusions in a logical step-by-step fashion. Turing showed the equivalence between a formal logical system and a Turing machine. In short, given any digital computer *C* with unlimited memory, we can find a formal system *F* such that the possible outputs of *C* coincide with the possible theorems of *F*, and conversely. Using this equivalence, Turing restated the Decision Problem in computer-theoretic terms as the Halting Problem considered earlier. And since the two problems are logically equivalent, the fact that the Halting Problem has no solution implies that the Decision Problem is also unsolvable. So again we run up against a brick wall in trying to get to the heart of things by following a set of rules.

Despite the pioneering work of Turing, which was carried out in the latter half of the 1930s, it was really the work of Kurt Gödel a few years earlier that applied the coup de grace to the idea that there was no difference that mattered between the real-world notion of a truth and the formal-system concept of proof. So now let's turn to a consideration of what Gödel did and what it means for the hope of understanding the world by following rules.

TRUTH IS STRANGER THAN PROOF

In his book *Infinity and the Mind,* mathematician and science-fiction writer Rudy Rucker recounts his visit to a church in Rome outside of which stands a huge stone disc. Carved onto this disc is the face of a hairy, bearded man whose slot-shaped mouth is located about waist level. According to popular legend, God has decreed that anyone who sticks their hand into the mouth and then utters a false statement will never be able to pull their hand back out again. Rucker states that he went to this church, stuck his hand into the mouth and said, "I will not be able to pull my hand back out again." Needless to say, Rucker left Rome with all appendages intact. This story illustrates the logical basis for why it will never be possible to produce a "universal truth machine" capable of generating all possible real-world truths.

Suppose such a universal truth machine (UTM) indeed does exists. Now feed the following statement *S* into the machine: "The UTM will never print out this statement." If the UTM ever does print out *S*, then *S* will be false. So the UTM will have printed out a false statement. But this is impossible since, by assumption, the UTM is supposed to print out only true statements. Therefore, the UTM will never print out *S*, implying that *S* is indeed a true statement. But now we have a true statement that the UTM will never print out, contradicting the fact that the machine is universal (i.e., that it will print out *all* true statements). The perceptive reader will recognize the self-referential sentence *S* as another version of the famous Epimenides paradox discussed in the opening section of this chapter.

The punch line of the preceding argument is that truth is not finitely describable. Perhaps it's not too surprising that there's no set of rules sufficient to generate all possible real-world truths. After all, the rules themselves exist within the world they are supposed to describe. So it's a little bit like pulling yourself up by your own bootstraps to ask for a finite set of rules that will generate the infinity of all possible real-world truths. What is surprising, however, is Gödel's proof that this same limitation holds for the much smaller and far less cluttered world of the whole numbers. Here's a brief outline of how Gödel came to this astonishing conclusion.

Usually the alphabet of a formal system will include a symbol that can be used to signify the idea of negation. We then call the system *consistent* if a statement *S* and its negation are not both theorems of the system. It's fairly natural to impose the requirement that a system be consistent, since in practice a formal system is generally introduced as a way of summarizing a collection of factual statements about some possible world. And there is no world in which a statement and its negation can both be true facts.

Gödel's results apply to all formal systems that are (1) finitely describable, (2) consistent and (3) rich enough to allow us to make all possible statements about the relationship among the whole numbers. For such systems, Gödel showed that there are two inherent limitations, called Gödel's First and Second Incompleteness Theorems.

The First Incompleteness Theorem says that no formal system is capable of deciding every statement about numbers. This means that for any formal system there exists a statement about the natural numbers such that neither the statement nor its negation is a theorem of the system. Thus, we have

> **Gödel's First Incompleteness Theorem—Formal Logic Version:** *For every consistent formalization of arithmetic, there exist arithmetic truths unprovable within that formal system.*

Remember that for a given mathematical structure like arithmetic, there are an infinite number of ways we can choose a finite set of axioms and rules of inference in an attempt to mirror syntactically the mathematical truths of the structure. What Gödel's result says is that *none* of these choices will work. In short, there are no rules for generating *all* the truths about the natural numbers.

Gödel's result is shown graphically in Figure 4.6 for a given formal system **M**, where each point of the square represents a possible statement about the relationship among numbers (e.g., statements like "The sum of two odd numbers is even" or "The square of every number is negative"). Some of these statements are true and others are false. We would like to use our formal system to show that all the true

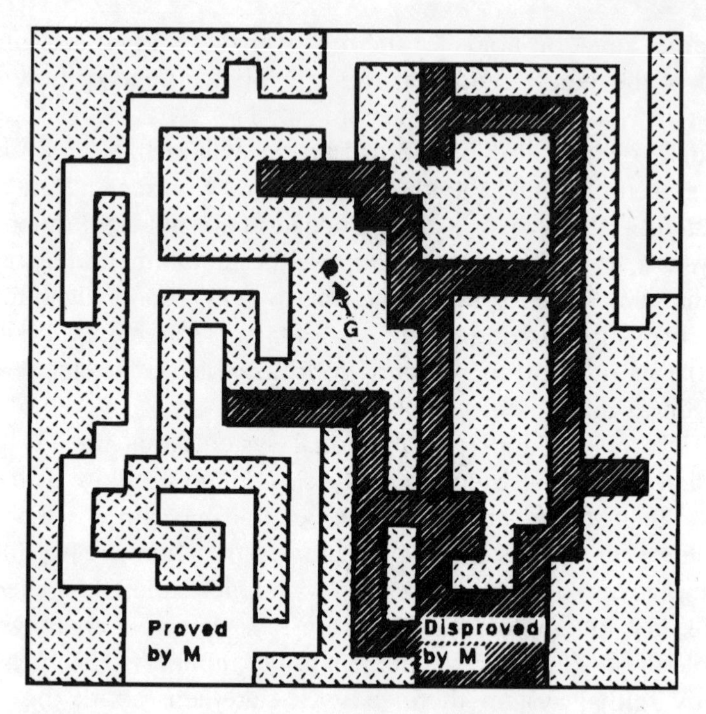

FIGURE 4.6. GÖDEL'S FIRST INCOMPLETENESS THEOREM IN LOGIC SPACE

statements correspond to theorems, while none of false ones do. Initially the square is colored entirely in gray. As we prove an arithmetical statement true using the rules of the formal system **M**, we color that statement white; if we prove the statement false, we color it black. Gödel's First Incompleteness Theorem (henceforth known more compactly as Gödel's Theorem) says that there will always exist statements like **G** that are eternally doomed to a life of gray; it's impossible to eliminate the gray and color the square entirely in black and white. And this result holds for *every* possible formal system **M**, provided only that the system is consistent. So, for every consistent formal system **M**, there is at least one statement **G** that can be seen to be true. Yet **G** is not a theorem of **M**. As with the rest of life, so it is too with arithmetic: there's no washing away the gray!

We call a statement like **G** *undecidable* since it can be neither proved nor disproved within the framework of the formal system **M**. And if we add that undecidable statement **G** as an axiom, thereby creating a new formal system, the new system will have its own *Gödel sentence* **G**—unprovable, yet true.

The general idea of the proof of this stunning result follows the lines outlined above for the Universal Truth Machine. Using the machine–formal-system equivalence, let $\mathbf{C_M}$ be the theorem-listing machine that prints out all the theorems of **M**. Suppose **G** is the statement "$\mathbf{C_M}$ will never print out this sentence." As we've already seen, it must be the case that **G** is a true sentence that will never be printed by $\mathbf{C_M}$.

The difficult part of the proof is to show that **G** can be put into the form of a statement about whole numbers. Gödel showed how to do this by a tricky process called Gödel numbering. The details of how this goes can be found in the material cited in the To Dig Deeper section for this chapter. When all the smoke clears away, the self-referential statement **G**, saying "**M** cannot prove **G**," has been transformed into a statement about whole numbers. More specifically, the arithmetic version of the statement **G** asserts that a certain polynomial equation has no integer solutions.

Given the equivalence between Turing machines and formal systems, it should come as no surprise to find that Gödel's Theorem has a computer-theoretic counterpart. In fact, this counterpart is nothing other than the by now familiar Halting Problem. So again we are led to the same sad conclusion; namely, that there are inherent, irremovable limitations on how far we can go toward "computing" the truth by following a set of rules.

Gödel's Second Incompleteness Theorem addresses the matter of consistency. Recall that the consistency of the system **M** can be stated in plain English as "There is no statement about numbers that is both true and false by the rules of **M**." Again using Gödel numbering, Gödel showed how to transform this statement into a purely arithmetical assertion involving the unsolvability of a certain polynomial equation. The Second Incompleteness Theorem then states that the formal system **M** is too weak to prove its own consistency.

The proof of the Second Incompleteness Theorem makes use of the proof of the First Incompleteness Theorem. At the end of the day, we manage to prove that the consistency of **M** implies the truth of the Gödel sentence **G**. But since we cannot prove **G** within **M**, it must also be the case that **M** cannot prove its own consistency. Table 4.4 summarizes the various steps in Gödel's road to incompleteness.

Gödel's Incompleteness Theorems give the appearance of settling once and for all the matter of generating truth from a set of rules. And

TABLE 4.4.
THE MAIN STEPS IN GÖDEL'S PROOF

Gödel Numbering: Development of a coding scheme to translate every logical formula in *Principia Mathematica* into a "mirror-image" statement about the natural numbers.

⇓

Epimenides Paradox: Replace the notion of truth by that of provability, thereby translating the Epimenides paradox into the assertion "This statement is unprovable."

⇓

Gödel Sentence: Show that the sentence "This statement is unprovable" has an arithmetical counterpart, its Gödel sentence **G**, in every conceivable formalization of arithmetic.

⇓

Incompleteness: Prove that the Gödel sentence **G** must be true if the formal system is consistent.

⇓

No Escape Clause: Prove that even if additional axioms are added to form a new system in which **G** is provable, the new system with the additional axioms will have its own unprovable Gödel sentence.

⇓

Consistency: Construct an arithmetical statement asserting that "arithmetic is consistent." Prove that this arithmetical statement is not provable, thus showing that arithmetic *as a formal system* is too weak to prove its own consistency.

so they do—to a point. But less than fifty years after Gödel published his work, the editors of the *Los Angeles Times* noted an extension of Gödel's results that was so startling in its implications that they were moved to comment in their editorial of June 18, 1988 that this result "makes the world shake just a little." What kind of mathematical result could possibly send the general press into such a state? Nothing less than a proof that the structure of arithmetic itself is random. I hasten to point out that this does *not* mean that 2 + 2 is usually 4, occasionally 3.9 and every now and then 17. Rather, it means that there are perfectly well-defined assertions about the integers whose truth or falsity is forever beyond the bounds of rational thought to decide. So in deciding what's what about such statements, there is no more systematic procedure than to toss a coin. While to examine this result would take us a bit too far off the track we are exploring here, the interested reader can find a complete account in the material cited in the To Dig Deeper section. What does lie right in the center of our path is the underlying basis of this shocking result—the concept of the complexity of a number.

OUT-GÖDELING GÖDEL

In the mid 1960s, Gregory Chaitin was enrolled in a computer-programming course being given at Columbia University for bright high-school students. At each lecture the professor would assign the class an exercise whose solution required the writing of a program. The students then competed among themselves to see who could write the shortest program that solved the given problem. While this spirit of competition no doubt added some spice to what were probably rather dull programming exercises, Chaitin reports that no one in the class could even begin to think of how to actually prove that the weekly winner's program was really the shortest possible.

Even after the course ended Chaitin continued pondering this shortest-program puzzle, eventually seeing how to relate it to a different question: how can we measure the complexity of a number? Is there any way that we can objectively claim that π is more complex than, say, $\sqrt{2}$ or 759? Chaitin's answer to this question ultimately led him to one of the most surprising and startling mathematical results of recent times.

In 1965, now an undergraduate at the City University of New York, Chaitin arrived at the bright idea of defining the complexity of a number as the length of the shortest program that will cause a Turing machine to print out the number. As it turns out, this concept of complexity is intimately related to Gödel's results.

Using the notion of the shortest program that will print out a given number as a way to characterize the complexity of that number, we can get a precise mathematical handle on the idea of a random, or incompressible, number. We call a number *random* if there is no program for calculating it whose length is shorter than the length of the number itself. Expressed another way, a number is random if it is maximally complex. Here, of course, we take the length of a number to be the number of bits (digits) in its binary expression.

But do random numbers really exist? The surprising fact is that almost all numbers are random! The reason is exactly the same as why almost all real numbers are uncomputable. To see this result, let's compute the fraction of numbers of length n having complexity less than, say, $n-5$. There are at most $1+2+\ldots+2^{n-5}=2^{n-4}-1$ programs of length at most $n-5$. Consequently, there are at most this many numbers of length n having complexity less than or equal to $n-5$. But there are a total of 2^n binary numbers of length n. Thus, the proportion of these numbers having complexity no greater than $n-5$ is at most $(2^{n-4}-1)/2^n \leq \frac{1}{16}$. So we see that less than one number in sixteen can be described by a program whose length is at least five digits less than the length of the number. Similarly, less than one number in five hundred has a length ten or more greater than its shortest program (i.e., its complexity). Using this kind of reasoning and letting $n \rightarrow \infty$, it's fairly easy to prove that the set of numbers with complexity less than their length forms an infinitesimally small subset of the set of all real numbers. Basically, there are too many real numbers and not enough Turing machine programs; almost every number is of maximal complexity (i.e., random). Now let's get back to the problem of shortest programs.

The starting point for Chaitin's remarkable results is the seemingly innocent query "What is the smallest number that cannot be expressed in a finite number of words?" This statement seems to pick out a definite number. Let's call it U for "unnameable." But thinking about things for a moment, we see that there appears to be something fishy about this labeling. On the one hand, we seem to have just described the number U in words. But U is supposed to be the first natural number that *cannot* be described in words. It's important to note here that describing something and computing it are two completely different matters. For example, it's easy enough to describe a unicorn, but it's something else again to write a program for making one. This description paradox, first suggested to Bertrand Russell by a certain Mr. Berry, a Cambridge University librarian, plays the same role in Chaitin's thinking about the complexity of numbers and programs as the Epimenides Paradox played in Gödel's work on the limitations of formal systems.

Recall that to bypass the issue of formalizing truth, Gödel had to substitute a related notion, provability, and talk about a statement being unprovable within a given formal system. Similarly, the Berry

Paradox contains its own unformalizable notion, the concept of denotation between the terms in its statement and numbers. Part of Chaitin's insight was to see that the way around this obstacle was to shift attention to the phrase "the smallest number not computable by a program of complexity n." This phrase *can* be formalized, specifying a certain computer program for searching out such a number. What Chaitin discovered was that no program of complexity n can ever produce a number having complexity greater than n. Therefore, the program of complexity n can never halt by outputting the number specified by Chaitin's phrase.

More generally, this result shows that even though there clearly exist numbers of all levels of complexity, it's impossible to prove this fact. That is, given any computer program, there always exist numbers having complexity greater than that program can recognize (i.e., generate). In the words of Georgia Tech physicist Joseph Ford, "A 10-pound theory can no more generate a 20-pound theorem than a 100-pound pregnant woman can birth a 200-pound child." Speaking somewhat informally, Chaitin's Theorem says that no program can calculate a number more complex than itself. The To Dig Deeper section outlines the proof of this very fundamental result linking computing, complexity and information.

The implication of Chaitin's result is that for sufficiently large numbers N, it cannot be proved that a particular string has complexity greater than N. Equivalently, a number N exists such that no number whose binary string is of length greater than N can be proved to be random. Nevertheless, we know that almost every number is maximally complex, hence random. We just can't prove that any *given* number is random. The problem is that each digit in a random sequence carries positive information since it cannot be predicted from its predecessors. Thus an infinite random sequence contains more information than all our finite human systems of logic put together. Hence verifying the randomness of such a sequence lies beyond the powers of constructive proof. Looking at the problem in another way, to write down an arbitrarily long string requires that we give a general rule for the entries of the string. But then this rule is shorter than suitably large sections of the string. So the string can't be random, after all! As one might suspect by now, this result is deeply intertwined with the other decision problems considered earlier.

To make this connection, consider a formal system whose axioms can be expressed in a binary string of length N. Chaitin's Theorem then says that there is a program of size N that does not halt—but we cannot prove that fact within this axiomatic system. On the other hand, this system can allow us to determine exactly which programs of size less than N halt and which do not. So we have another way of proving the undecidability of the Halting Problem, since the complexity results state that there always exist programs for which we cannot determine in advance whether or not they will stop. And, of course, Chaitin's Theorem offers another perspective on Gödel:

> **Gödel's Theorem—Complexity Version** *There exist numbers having complexity greater than any theory of mathematics can prove.*

So if we have some theory of mathematics (i.e., a formal system), there always exists a number t such that our theory cannot prove that there are numbers having complexity greater than t. Nevertheless, by jumping outside the system ("jootsing"), we can clearly see that such strings exist. To generate one, simply toss a coin a bit more than t times, writing down a 1 when a head turns up and a 0 for tails.

It's thought-provoking to consider the degree to which Chaitin's result imposes limitations on our knowledge about the world. Rudy Rucker has made the following estimate: suppose K represents our best present-day knowledge, while C denotes a universal Turing machine whose reasoning powers equal those of the smartest and cleverest of human beings (we'll take up the existence of such a UTM in the next section). Rucker then estimates the maximal complexity number t in Chaitin's Theorem to be

$$t = \text{complexity } K + \text{complexity } C + 1 \text{ billion}$$

where the last term is thrown in to account for the overhead in the program of the machine C. Plugging some plausible numbers for K and C into this expression, Rucker concludes that t is less than 3 billion. So if any worldly phenomenon generates observational data having complexity greater than around 3 billion, no such machine C (read: human) will be able to prove that there is some short program (i.e., theory) explaining that phenomenon. Thus, appealing to the idea of scientific theories as tools for reducing the arbitrariness in observations, Chaitin's work says that our scientific theories are basically powerless to say anything about phenomena whose complexity is much greater than about 3 billion. This does *not* mean that there is no simple explanation for these phenomena. Rather, it means that we will never understand this "simple" explanation—it's too complex for us! Complexity 3 billion represents the outer limits to the powers of human reasoning. Beyond that we enter the "twilight zone" where reason and systematic analysis give way to intuition, insight, feelings, hunches and just plain dumb luck.

By now we've seen ample evidence of the equivalence between a formal system of logic and a computer program. But there is yet another equivalence between these two objects that bears directly on our earlier discussions of dynamical systems and chaos in Chapter Three. That is the equivalence between a computer program and a dynamical system. Table 4.5 spells out the dictionary establishing that the two are mirror images of each other.

If there's any message for mankind at all in these results of Gödel, Turing and Chaitin, it's that there is a forever unbridgeable gap between what's true and what can be proved. So where do chaos and strange attractors fit into the overall scheme of things? Appealing to the dictionary of Table 4.5, we can argue that the existence of chaotic dynamical processes forms a natural link between Chaitin's complexity results and Gödel's Theorem. Moreover, the existence of a rich variety of real-world truths that we can know for sure depends in an essential

TABLE 4.5.
THE DYNAMICAL SYSTEM–COMPUTER PROGRAM DICTIONARY

Dynamical System	*Program*
Number field	Tape symbols
State manifold	All possible tape patterns
State	A tape pattern
Constraints	Set of admissible tape patterns
Initial state	Input tape pattern
Vector field	Program instructions
Trajectory	A sequence of tape patterns
Attractor	Tape pattern when the program halts or goes into an infinite loop

way upon the existence of such strange attractors. Here's the line of argument supporting this claim.

We have seen that the theorems of a formal system, the output of a UTM and the attractor set of a dynamical process are completely equivalent; given one of these objects, it can be faithfully translated into either of the others. But the idea of a provable real-world truth coincides with the decoding of a theorem in a formal system. Therefore, let's employ the symbol T to represent the universe of true statements, while using P to denote the set of theorems provable in some formal system. Of course, Gödel's Theorem states that under a given interpretation, the set of provable theorems P is only a proper subset of the set of true statements T.

From the discussion on complexity, we saw that there exist computable numbers of arbitrary complexity. Each of these computable quantities corresponds to the attractor of some dynamical system. Consequently, since there are an infinite number of strings on the attractor of these computable strings (one string for each digit in the real number that the attractor corresponds to), there necessarily exist dynamical systems whose attractor set is infinite. But fixed points and limit cycles are both attractors having only a finite number of strings. Hence there must exist something "bigger." But Chaitin's Theorem tells us that the attractor set must be smaller than the whole state manifold since it asserts that there are strings that can never be computed. In short, Chaitin's Theorem implies the existence of some kind of attractor beyond a fixed point or a limit cycle. This can only be a strange attractor.

Now let's assume that such strange attractors exist. Since they do not fill up the entire state manifold, there must exist states that cannot be reached from any given initial state on the attractor. But from the equivalence of formal systems and dynamical systems, this is just another way of saying that P is a subset of T. This relationship, of course, is just a restatement of Gödel's Theorem. Putting these two sets of arguments together, we have the chain of implications:

$$\text{Chaitin's Theorem} \Rightarrow \text{strange attractors} \Rightarrow \text{Gödel's Theorem}$$

As the pièce de résistance of our tour, we come to the perhaps surprising fact that chaos implies truth, in the sense that a world without strange attractors and, hence, without chaos would be very impoverished in the number of mathematical theorems that could be proved. This conclusion, in turn, implies that whatever real-world truths might exist, the overwhelming majority of them cannot be the counterparts of theorems in any formal logical system. Of course from this perspective we might already be living in such a world. But the existence of strange attractors allows us to hold out the hope that the gap between proof and truth can at least be narrowed—even if it can never be completely closed.

Before leaving this discussion of dynamical systems and computing machines, it's worth a moment to consider what it is exactly about computing machines that gives rise to things like the unsolvability of the Halting Problem and uncomputable numbers. Basically, the problem is the assumption that it takes a *fixed*, finite amount of time to carry out a single step in a computation. For his idealized computer, Turing assumed an infinite amount of memory. Mathematician Ian Stewart, on the other hand, considers the Rapidly Accelerating Computer (RAC), whose clock accelerates exponentially fast, with pulses separated by intervals of ½, ¼, ⅛ . . . seconds. So the RAC can cram an infinite number of computational steps into a single second. Such a machine would be a sight to behold as it would be totally indifferent to the algorithmic complexity of any problem presented to it. On the RAC, everything runs in *bounded* time.

The RAC can calculate the incalculable. For instance, it could easily solve the Halting Problem by running a computation in accelerated time and throwing a switch if and only if the program stops. Since the entire procedure could be carried out in no more than one second, we then only have to examine the switch to see if it's been thrown. The RAC could also prove or disprove famous mathematical puzzles like Goldbach's Conjecture (every even number greater than 2 is the sum of two primes). What's even more impressive, the machine could prove all possible theorems by running through every logically valid chain of deduction from the axioms of set theory. And if we believe in classical Newtonian mechanics, there's not even a theoretical obstacle in the path of actually building the RAC.

In Newton's world, we could model the RAC by a classical dynamical system involving a collection of interacting particles. One way to do this, suggested by Z. Xia and J. Gerver, is to have the inner workings of the machine carried out by ball bearings that speed up exponentially. Because classical mechanics posits no upper limit on the velocities of such point particles, it's possible to accelerate time in the equations of motion by simply reparameterizing it so that infinite subjective time passes within a finite amount of objective time. What we end up with is a system of classical dynamical equations that mimics the operations of the RAC. Thus, such a system can compute the uncomputable and decide the undecidable.

Of course, just like Turing's infinite-memory machine, the RAC is impossible in the real world. The problem is that at the nitty-gritty level of real material objects like logical gates and integrated circuits, there is a theoretical upper bound to the rate of information transfer (i.e., velocities). As Einstein showed, no material object can exceed the velocity of light. Thus, there is no RAC and, hence, no devices to complete the incompletable.

The fact that machines are limited in their ability to uncover the last secret of the cosmos has not dampened the enthusiasm of many scientists and philosophers for arguing that human beings are also in some sense machines. And from this assumption it's but a short step to the conjecture that there is no barrier in principle to constructing a living, thinking being in plastic, copper and silicon. What's meant by this is an entity that is genuinely alive and capable of cognitive thought in just the same sense as our fellow men and women are distinguishable from cars, bacteria, stones and other objects that are not alive and cognitively aware. Let's use some of our results about rationality and mechanism to examine this wildly ambitious claim. But before doing so, a quick summary of the situation seems in order.

The results of Gödel and Chaitin showed that we'll never get at all the truth by following rules; there's always something out there in the real world that resists being fenced in by a deductive argument. Nevertheless, aficionados of algorithms believe that rules are enough to duplicate both the cognitive and the material functioning of human beings. This can only mean that either (1) there is some way for humans to escape the Gödelian net or (2) humans are just not as special as we'd like to believe. The remainder of the chapter looks at these two possibilities.

REAL BRAINS, ARTIFICIAL MINDS

Boston University computer scientist Ed Fredkin likes to tell the parable of a computer simulation he calls the Heaven Machine. It goes like this. One day you read an advertisement from the Heaven Machine Corporation, offering you the opportunity to have an exact copy of your brain states loaded into a gigantic computer simulation. If you accept the company's offer, however, the duplication process destroys your original brain, at which point your life on Earth is unfortunately over. By way of compensation, though, the Heaven Machine Company promises you eternal life within the machine. Further, they claim that this life in the machine is nothing short of heavenly.

Because you look pretty skeptical about the whole business, the HMC salesman offers to let you talk the matter over with Joe, your neighbor who recently signed up with the company and had his brain duplicated in the machine. So they take you into a room with a huge computer screen, which initially is pretty fuzzy, but after a while the picture comes into focus, and you see your neighbor. "Hi there, Joe," you say. "What's happening?" Joe replies, "Life is fantastic. Everything is just heavenly up here. You wouldn't believe some of the conversations I've had lately. There are so many amazing people to talk with—Aristotle, Newton, Buddha. And when I'm not trading ideas with these guys, I play a lot of golf, sun myself on the beach and, in general, relax and enjoy all the things I always wanted to do but never had the time for when I lived next door to you. And the social life! My datebook looks like a dentist's calendar. It's enough to make Warren Beatty look like a fumbling teenager. Believe me when I tell you it's like I died and went to heaven." Hearing all this you think, that's my old pal Joe, all right. No question about that!

So is this a genuine heaven-on-Earth machine or just a nightmarish fantasy? Is it even remotely plausible that by running a computer program—by following a mere set of rules—it would be possible to capture the essence of what it means to be Joe? Or me? Or you? While it's far from apparent on the surface, this query is only a special case of the more general Big Question that we've been leading up to throughout this chapter: Is every observable real-world phenomenon just the result of following a simple set of rules? Put more prosaically, is the universe itself just one big computer?

In 1950, Alan Turing published the paper "Computing Machinery and Intelligence," which sparked off a debate that rages to this day over the question Can a machine think? In addition to its pivotal role in drawing attention to the matter of machine intelligence, Turing's paper was notable for its introduction of an operational test for deciding whether or not a machine really was thinking—human style. This criterion, now termed the *Turing test*, is unabashedly behavioristic in nature, involving the machine's fooling a human interrogator into thinking it is actually a human. Turing's rationale for proposing what he called the Imitation Game was that the only way we have for deciding whether or not other humans are thinking is to observe their behavior. And if this criterion is good enough to decide if humans are thinking, then fairness to machines dictates that we should apply the same criterion to them.

On November 8, 1991 the Boston Computer Museum held the world's first hands-on Turing test, in which eight programs conversed with human inquisitors on a restricted range of topics that included women's clothing, romantic relationships and Burgundy wine. At the day's end, the judges awarded first prize to a program called *PC Therapist III*, which was designed to engage its questioner in a whimsical conversation about nothing in particular. For example, at one point the program suggested to a judge, "Perhaps you're not getting enough affection from your partner in the relationship." The judge replied, "What are the key elements that are important in relationships in order to prevent conflict or problems?" "I think you don't think I think," responded the machine.

This kind of interchange did little to fool the judges, most of whom said they were able to spot the mistakes, rooted in a lack of everyday common sense, that immediately singled out the computer programs from the humans. Nevertheless, the overall conclusion from this historic experiment was that perhaps the Turing test isn't as difficult as many people originally thought, since even the primitive programs in this contest managed to fool some of the judges most of the time. Of course, we should keep in mind that this wasn't a *true* Turing test, since the domains of discourse were severely restricted. But it was still a pretty good initial approximation.

A telling argument against the adequacy of the Turing test as a benchmark of intelligence has been advanced by philosopher Ned Block. Suppose, he argues, that we write down a tree structure in which every possible conversation of less than five hours' duration is explicitly mapped out. This structure would clearly be enormous, much larger than any existing computer could store. But for the sake of argument, let's ignore this difficulty.

By following this tree structure, the machine would interact with its interrogator in a way indistinguishable from the way an intelligent human being would do so. Yet the machine would simply make its way through this tree, which strongly suggests that the machine has no mental states at all. And this same conclusion holds for any conversation of finite duration.

From this argument, Block draws the moral that thinking is not fully captured by the Turing test. What's wrong with the tree structure is not the behaviors it produces but the *way* it produces them. Intelligence is not just the ability to answer questions in a manner indistinguishable from that of an intelligent person; to call a behavior intelligent is to make a statement about how that behavior is produced.

It's clear that the Turing test represents a third-party perspective on human intelligence. Standing outside the system, the test is designed to discern human intelligence in a machine by observing only the behavioral output of the machine. The Turing test says nothing about the internal constitution of the machine, how its program is structured, the architecture of the processing unit or its material composition. In Turing's view of intelligence, only behavior counts. And if you have the "right stuff," then you are a thinking machine.

In 1989, theoretical physicist Roger Penrose published *The Emperor's New Mind*, a book whose central argument is that the human mind is capable of transcending rational thought, hence can never be duplicated in a machine. Before going on, let me note that we are using the term *rational thought* in the strong sense of following rules or an algorithm to arrive at a result by a process of logical inference. There is no connection here with the everyday economic interpretation of rationality as relating to self-interest or prudent action. Penrose's message, which he justified by a wildly speculative appeal to quantum processes in the human brain as the basis of consciousness and intelligence, was a great source of comfort, I'm sure, to many computerphobes and other anti–artificial intelligence types, a fact that doubtless goes a long way toward explaining the presence of such a technical book on the best-seller lists for months on end. Be that as it may, Penrose's anti-AI argument comes down to the claim that the human mind is somehow bigger than rational thought.

A key ingredient in Penrose's argument is Gödel's famous result showing that there are true statements of arithmetic that the human mind can know but that cannot be the end result of following a fixed set of rules (i.e., a computer program). While there are well-known reasons why Gödel's Theorem should be regarded skeptically when it comes to using it as an argument against thinking machines, for our purposes here what's important about Gödel's result is that it suggests there are limits to the rational powers of the human mind. The big question for mechanists then becomes Can these limits be removed, or at least extended? As a prologue to confronting this issue head-on, let me first outline briefly the basic positions today of both the pro- and anti-AI camps. For more details on these competing schools of thought, the reader is invited to consult the general accounts, as well as the more specialized discussions, cited in the To Dig Deeper section for this chapter.

Pro-AI

As a rough classification, the pro-AI world is divided into two basic schools: Top Down and Bottom Up. The first sees the basic hardware of the brain as irrelevant to the issue of duplicating human intelligence in a computing machine. Consequently, Top Down attempts to capture what the brain does center upon trying to extract the rules that the brain uses and then coding these rules in a form congenial to computing machines.

Bottom Uppers, on the other hand, argue that perhaps the way our particular human type of brain is physically constituted plays a crucial role in our cognitive abilities. If so, the argument goes, then it's impossible to capture cognition—human style—in a machine without respecting this physical structure. Thus, these so-called New Connectionists focus their attempts to mimic the mind in a machine on constructing programs whose functional organization mirrors as closely as possible that of the human brain. Now by way of prelude to consideration of the "minds as machines" problem, let me take a longer look at the details of these two very different approaches to the problem of capturing human thought in a machine. First, the Top Downer's view of the world.

The key words in the Top Down vocabulary are *representations* and *rules*. Since the time of the very first Top Down program—the General Problem Solver created by Herbert Simon, Alan Newell and

Cliff Shaw in the 1950s—the twin problems confronting these research efforts have revolved about how knowledge is to be represented in symbols and what rules should be used to combine these symbol strings into new, cognitively meaningful strings. This description makes it clear, I think, that the research manifesto of the Top Down view of AI has no place in it for the actual neurophysiological hardware of a real brain. Rather, the time and energy of Top Down researchers is devoted to a search for clever representation schemes and what might be called the "rules of thought." In short, these research agendas are focused on skimming off the symbolic representation schemes and rules of thought used by the brain, ignoring totally the brain's actual hardware. This line of investigation has been divided by the Top Downers' archenemies Hubert and Stuart Dreyfus into three rather distinct phases:

• *Representation and search (1955–1965)*—During this period work centered on showing how a computer can solve certain classes of problems using the general heuristic search technique termed *means-end analysis.* This involves making use of any available operation that reduces the distance between the current state of the system and the description of the desired goal. Simon and Newell made extensive use of these ideas, abstracting the heuristic technique for incorporation into their General Problem Solver.

• *Microworlds (1965–1975)*—Early on, the Top Down approach scored some seemingly impressive victories, especially in severely restricted areas like geometric theorem-proving, chess-playing and other areas in which problems could be solved with the combination of a large amount of formal logical manipulations and a minimal amount of real-world background knowledge. Unfortunately, it soon became clear that most everyday human problem-solving did not involve problems with this happy conjunction of features. Experience with language translation by machine brought out especially clearly the fact that most human cognition involves a considerable amount of background knowledge about the world, what many have termed "tacit knowledge." For example, one Russian-English translation program translated the English idiomatic phrase "out of sight, out of mind" into the Russian equivalent of "blind and insane." With these kinds of problems emerging at an ever-increasing rate as researchers tried to create programs for practical, everyday tasks, the question for Top Downers became how to account for the necessary background knowledge in their rules and representations.

One early attempt was a kind of AI version of the age-old Procrustean fit. Namely, create an artificial world inside the machine, a world about which the machine has complete knowledge. Of course, these artificial worlds or, as they came to be termed, *microworlds*, are vastly slimmed-down versions of the real thing. But the hope was that by abstracting those features of the real world deemed important for a given task, the machine could then be given enough background information to be able to think intelligently about objects and relationships in these scaled-down, fake worlds.

An early example of this artificial worlds approach to AI was Terry Winograd's *SHRDLU*, a microworld consisting of simulated three-dimensional objects like blocks, pyramids and spheres of different colors. The computer was given all the relevant information about the properties of these objects, after which instructions would be given by the operator, telling the machine to perform certain operations on these objects. A typical instruction might be something like "Pick up the blue block and place it on top of the red sphere." Knowing the properties of blocks and spheres, the computer might then respond that it couldn't carry out the order since a square block cannot sit stably atop a sphere.

The hope was that these microworlds could be gradually made more realistic and combined with other microworlds so as to approach real-world understanding. Although these kinds of efforts yielded some limited successes, it soon became evident that the whole research program was based upon a crucial misunderstanding—of the difference between a *universe* and a *world*. Thinking of things like the *world of business* or the *world of science* shows that a world is an organized body of objects, purposes, facts, skills and practices that give meaning to human activities. On the other hand, a set of interrelated facts can constitute a universe without being a world. The difference becomes apparent by considering the meaningless *physical universe* and the meaningful *world of physics*. Unfortunately, microworlds are not worlds but isolated, meaningless domains. And it gradually became evident that there was no way such domains could be combined and extended to encompass the many worlds of daily life.

• *Commonsense knowledge (1975–present)*—The first two periods of Top Down AI were characterized by efforts aimed at seeing how much could be done with as little knowledge as possible. But the commonsense-knowledge problem could not be swept under the rug forever. The obvious next step was to try to introduce data structures for stereotyped situations as a way to incorporate everyday, taken-for-granted knowledge into computer programs. But with the failure of approaches like Marvin Minsky's "frames" and Roger Schank's "scripts," it finally became clear that a radically new approach was needed. It was time to give up on the conviction of Descartes and the early Wittgenstein that the only way to produce intelligent behavior is to mirror the world with a formal theory in the mind. At this point, classical Top Down, symbol-based AI became an example of what Imré Lakatos has called a "degenerating research program." Enter the Bottom Up view of the world.

Looking at the brain from the other end of the telescope, Bottom Up proponents argue that the physical hardware of the brain—or at least its general architecture—does matter when it comes to human cognition. And if we're to have any hope of duplicating that kind of intelligence in a machine, it behooves us to explicitly account for the organizational structure of that hardware in our programs. What this means is that we need to take an "insider's" perspective, looking at how the brain is wired and how that wiring serves to generate the kind of observable behavior we label *intelligent*.

Here are a handful of features that just about every neurophysiologist and connectionist agrees are characteristic aspects of the human brain:

• *Simple processors*—Most of the work of the brain is done by an unimaginably large number of neurons, each of which, taken by itself, can be regarded abstractly as an ultraprimitive computer, not much more complicated than a simple ON-OFF switch.

• *Massive parallelism*—The 10 billion or so neurons of the brain are connected via a network of dendrites, axons and synapses, resulting in concurrent operation of the neurons. Here the dendrites correspond to the neuron's input channels, while the axons are the outputs. The synapses can then be thought of as a kind of volume control mediating between the inputs and outputs.

So if we imagine each neuron as a light bulb, a motion picture of the brain in operation would show an array of billions of lights flashing on and off in a bewildering variety of patterns. This picture would look much the same as a Times Square message board, consisting of many rows of individual flashing lights that, taken together, form a recognizable pattern. The problem is that, at present, we haven't the foggiest idea of how to interpret these patterns.

- *Unprogrammed*—In contrast to a modern digital computer, whose program of instructions is inflexible, or "brittle," the brain seems to be relatively unprogrammed. The strengths of the synaptic connections, which determine the firing pattern of the neurons, seem to be governed more by various learning procedures in the brain than by any kind of explicit instructions laid down by a "programmer."
- *Adaptable*—The connective pattern in the brain is very plastic, allowing the brain to, in essence, reprogram itself by adjusting the synapses. This in turn provides the basis for things like memory, learning and creative thought. The necessity of plasticity for memory and learning is clear, I think, since both processes involve the brain's changing its state in some semipermanent fashion. Moreover, since by definition a creative thought represents something new that's generated and stored in the brain, it too must arise as a result of the brain's ability to reconfigure its neuronal connections somehow.

Using these desiderata as a checklist, Bottom Up proponents of AI have been busying themselves with the development of neural networks aimed roughly at mimicking in one way or another the overall architecture of the brain. The general structure of such a network is shown in Figure 4.7. The basic idea is to associate each processing element in the net with a physical neuron in the brain. The synaptic connections between these elements each have a numerical weight that can be either positive or negative. The size of the weight controls the influence that one element has on another, with a positive connection serving to excite an element and a negative connection inhibiting it. Overall, the activation of an element is determined by a combination of the excitatory and inhibitory influences it receives from its neighbors. Such a net is then trained to respond to a set of patterns by adjusting these weights. Once the net has been trained, the network is able to respond similarly (intelligently?) to other patterns bearing features in common with those present in the training set.

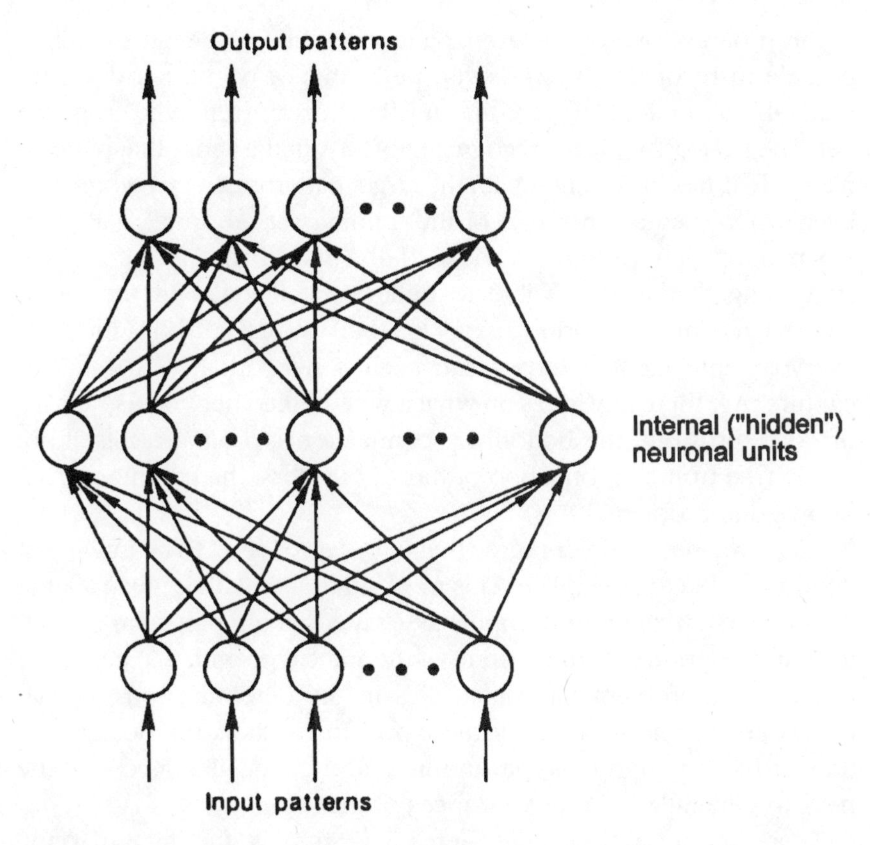

FIGURE 4.7. STRUCTURE OF A PROTOTYPICAL NEURAL NETWORK

So instead of dictating the rules of thought, a Bottom Up researcher sees whatever rules of thought the machine embodies as emerging from the reconfiguration of the pattern connecting the many processors as the machine learns to survive in its environment. In this essentially internal view of cognition, the machine starts out with very little knowledge of the world but with a hardware configuration that's pretty plastic. As time goes on and the machine interacts with its environment, certain patterns of thought prove more effective at solving the problems of daily life than others. These successful patterns then ultimately get wired into the connections linking the machine's processors. It is these connections that then determine whatever rules we can say the machine uses in arriving at its actions and decisions.

A neural network trained to respond to a set of patterns displays many of the features we associate with the human brain. For example, if only a part of one of the target patterns is shown, the net still might be able to recognize it. Moreover, performance of the net degrades gradually as individual neurons "misfire," or are removed from the net. The network can also recognize novelty, in the sense that patterns having features in common with the target patterns can be recognized, even when they are not part of the training set. All in all, a Bottom Up neural net represents a view of the world that is theory-free, suggesting that it is possible to behave intelligently in the world without having an a priori theory of that world. But how much of everyday intelligent behavior can such a network be expected to capture? Are there any limits on what a well-trained network is capable of? As it turns out, the Bottom Up connectionist program is itself not totally free from that bugaboo of the AI business, the commonsense-knowledge problem.

All neural net modelers agree that in order for a net to be intelligent it must be able to generalize. This means that if the net is given a large enough set of examples of inputs associated with a particular output, it should respond to further inputs of the same type with that particular output. The problem, of course, lies in the determination of what constitutes an input of "the same type." In practice, the designer of the net has a notion of "type" in mind, and counts it a success if the net can generalize to other instances of this class.

There are two difficulties here. The first is the fact that the boundaries separating classes of patterns is set *in advance* by the net's designer. Thus, the possibility for novel forms of behavior, which is clearly part of what we call human intelligence, is severely restricted. The second problem arises when the net produces an unexpected response to a given input. Can we really say then that the net has failed to generalize? Perhaps the net has been acting all along on a different idea of type, and this unexpected association has just revealed that difference.

In light of these difficulties, we might conclude that in order for a neural net to share our human sense of generalization, it must also share the size, connective structure and initial configuration of the brain. Moreover, it must share our idea of what constitutes an appropriate output, suggesting that it must share our experiences, needs, desires and emotions, as well as have a humanlike body with appropriate physical abilities for movement, sensory inputs and the like. If this indeed turns out to be the case, then the Bottom Up program will founder on exactly the same shoals as the Top Downers—the reef of commonsense knowledge. To quote Hubert and Stuart Dreyfus, "If the minimum unit of analysis is that of a whole organism geared into a whole cultural world, neural nets as well as symbolically programmed computers still have a very long way to go."

The one thing that both Top Down and Bottom Up devotees agree on is the claim that there is no obstacle in principle to the duplication of human cognitive capacity in a machine. The point of dispute is only how to go about doing it. But an impressive array of arguments have been marshaled against the very idea of such "strong AI." So let's now give the floor over to those adhering to the view that machines will never think like you and me.

Anti-AI

While the majority of AI advocates are computer scientists, psychologists, mathematicians and others of that ilk, most of the arguments against AI have been put forward by philosophers. Their outbursts against the very idea of strong AI seem to be based on one of three main lines of attack: phenomenology, antibehaviorism or Gödel's Theorem. Let me spend a page or two describing these counterarguments.

- *Phenomenology*—One of the most popular spokesmen for the anti-AI cause has been Hubert Dreyfus, a philosopher at the University of California at Berkeley. Dreyfus, along with his brother Stuart, a professor of engineering who's also at Berkeley, argues against the possibility of strong AI by appealing to the works of the phenomenological philosophers Martin Heidegger, Edmund Husserl and Maurice Merlau-Ponty. These giants of modern Continental philosophy claim that there are many human cognitive activities that simply cannot be thought of as the end result of the following of a set of rules. A favorite example of the Brothers Dreyfus in this regard involves learning how to drive an automobile.

According to the Dreyfuses, gaining expertise at driving a car involves a successive passage through five identifiable stages:

1. **Novice:** At this lowest skill level, context-free rules for good driving are acquired. Thus, one learns at what speed to shift gears and at what distance it's safe to follow another car at a given

speed. Such rules ignore context-sensitive features such as traffic density and weather conditions.

2. **Advanced beginner:** Through practical on-the-road experience, the novice driver learns to recognize concrete situations that cannot be described by an instructor in objective, context-free terms. For instance, the advanced beginning driver learns to use engine sounds as well as context-free speed as a guide for when to shift gears, and learns to distinguish the erratic behavior of a drunk driver from the impatient actions of an aggressive driver in a hurry.
3. **Competence:** The competent driver begins to superimpose an overall driving strategy upon the general rule-following behavior of the novice and the advanced beginner. He or she is no longer merely following rules that permit safe and courteous operation of the car, but drives with a goal in mind. To achieve this goal, the competent driver may now follow more closely than normal, drive faster than is allowed or in other ways depart from the fixed rules learned earlier.
4. **Proficiency:** At the previous levels, all decisions were made on the basis of deliberative, conscious choices. But the proficient driver goes one step further and makes decisions on the basis of a feel for the situation. There is no deliberation; things just happen. So, for example, the proficient driver when attempting to change lanes on a busy freeway may instinctively realize that there's another car coming up on the blind side and delay making a move. This instinctive reaction may arise out of memories of past experience in similar situations, although it may appear as an unexplainable lucky guess to an outside observer. Somehow there is a spontaneous understanding, or "seeing," of a plan or strategy.
5. **Expert:** An expert driver no longer sees driving as a sequence of problems to solve, nor does he or she worry about the future and devise plans. Such a driver becomes one with the car and has the experience of simply driving rather than driving a car. Thus, an expert driver has an intuitive understanding of what to do in a given setting. He or she doesn't solve problems and doesn't make decisions but just does what normally works.

The moral of this fable in five parts is that there is more to intelligence and expertise than mere calculative rationality. Expertise doesn't necessarily involve inference; the expert sees what to do *without* applying rules. This is the essence of the Dreyfus argument against the possibility of a rule-based program's ever achieving anything that remotely approximates genuine human intelligence.

I think it's clear from this example that the Dreyfus position against strong AI is essentially a third-person, "outside-the-system" argument against rule-based actions. The claim is that simply by looking at the external behavior of a human being, we can see cognitive activity that cannot be attributed to following a set of rules. The Dreyfuses then claim that it's impossible to program a computer to drive a car in a manner indistinguishable from an expert human driver, since the computer would have to have a set of rules to do such a task. But there are no such high-level rules that human drivers follow in guiding their vehicles through rush-hour congestion and high-speed freeway traffic. And since machines can only follow the rules encoded into their programs, the full experience of even such a relatively simple human task as driving a car cannot be captured within the confines of a computer program. Ergo, machines cannot think—at least not like you and me.

As an aside, it's of some significance here to note that Hubert Dreyfus admitted to me in a private conversation recently that the main thrust of his anti-AI argument is directed against the Top Down approach to strong AI, and that it may well be possible to duplicate human capacity in driving and everything else by following a Bottom Up approach. But even this approach may be stymied by the same commonsense-knowledge barrier already noted above. Now let's pass on to consideration of another class of reasons why machines will never think.

• *Antibehaviorism*—One of the strongest arguments yet launched by the philosophers against strong AI is that given by John Searle, also a professor of philosophy at Berkeley. Searle's argument is essentially a first-person, "inside-the-system" claim that what goes on within a computing machine when it moves symbolic representations around in accordance with a program is pure syntax. But, Searle argues, no amount of syntax alone (i.e., symbol shuffling) can ever give rise to semantics. In other words, the computer can have no understanding of the *meaning* of the symbols it manipulates. And without meaning there is no intelligence.

To dramatize this first-person point of view on thinking, Searle constructed the colorful analogy now called the Chinese Room argument. This involves imagining someone ignorant of Chinese being locked up in a room with a dictionary containing only Chinese ideographs. The room also contains a set of cards, each of which has a Chinese character printed on it. The person receives similar cards with Chinese characters through a slot in the door to the room. He or she then looks up the character on the card in the dictionary, passing back out through the slot the card containing the character called for by what's found in the dictionary listing.

It's clear that from the perspective of the person in the room, there is no understanding of Chinese here at all (i.e., there is no semantics). There is only a purely syntactic shuffling of cards back and forth through the slot in accordance with rules dictated by the dictionary. But from the third-person, essentially behaviorist perspective of a native Chinese speaker standing outside the room, the sequence of cards going in and out of the slot may well be seen as a perfectly sensible dialogue in Chinese about, say, tomorrow's weather, the state of the stock market or the end of the world. Searle's point here is that the actions of the person in the room duplicate exactly what happens inside a computer as it goes about its business of transforming one set of symbol strings into another.

When Searle first published this anti-AI argument in 1980, the howls of outrage from the pro-AI community could be heard from Stanford to MIT and back again. Here is just a small telegraphic sampling of the kinds of rebuttals offered against Searle's gedanken experiment.

Systems Reply: The essence of this rejoinder is to move the problem up to another level. While it's true that the person inside the Chinese Room doesn't understand the story, that person is merely a part of a whole system. And the system does understand the conversation. Thus, understanding is now ascribed to an overall system of which the person inside the room (i.e., the computer's central processor) is only a part.

The Brain Simulator: This reply involves imagining a program that simulates the sequence of neuron firings at the synapses of the brain of a Chinese speaker when understanding questions in Chinese and giving appropriate responses. Here the argument revolves around the claim that since the program is a perfect simulation of what's going on inside the brain of the Chinese speaker at the level of the synapses, either both the program and the native speaker understand the dialogue or they both don't. So what could be different about the program of the computer and the program of the Chinese brain?

Other Minds: In this case the argument is essentially that of the Turing test. How do we know that other people understand Chinese or anything else? Answer: Only by their behavior. But the person in the Chinese Room passes a behavioral test like a native Chinese speaker. So if you're going to attribute cognitive understanding to other people, you must also attribute it to a computer that can pass the same behavioral tests.

Several other lines of attack have also been leveled against the Chinese Room experiment. For Searle's response to these quibbles, and more details of the arguments themselves, the reader should consult the peer commentary accompanying the original article, which is cited in the To Dig Deeper section for this chapter. On this inconclusive note, let's move on to our final class of anti-AI claims, those resting on an appeal to Gödel's First Incompleteness Theorem.

• *Gödel's theorem*—One of the most influential arguments against the possibility of strong AI was advanced in 1961 by Oxford philosopher John Lucas, who appealed to Gödel's result saying, in effect, that since there exist arithmetical truths that we humans can see to be true but that a machine cannot prove, the capacity of the human mind must transcend that of any machine. As noted earlier, Roger Penrose, another Oxford don, recently appealed to much the same line of argument to also conclude that machines cannot think like humans. Penrose adds a twist to the usual Gödelian line by speculating that at least some part of human thought involves making contact with uncomputable quantities. And the best way he can explain how this might come about is by invoking mysterious quantum events influencing the brain's neuronal firing patterns.

For over thirty years now, the arguments have raged hot and heavy against these Gödelian arguments against strong AI, and I don't want to bore the reader by going into them again here. They can all be found in many places. Let it suffice to say that Gödel's theorem involves certain hypotheses, most importantly that the formal system (i.e., computer program) be logically consistent. It's rather clear, I think, that satisfaction of this condition by the human mind is a dubious assumption at best. I'm sure we can all remember instances when we behaved in a demonstrably inconsistent manner. And if the system is logically inconsistent, all bets are off as far as appeals to Gödel's result are concerned.

With all these arguments casting doubt on the feasibility of both the Top Down and Bottom Up programs for strong AI, is it really plausible that we'll ever be able to *design* a brain? After all, Gödel's result tells us that there are limits to what we can do by way of rationally planning anything. So perhaps both the symbol-processing and connectionist exercises are doomed to failure from the very outset. Maybe the brain is just too complex for us to ever fully understand, thus precluding our being able to design a machine to duplicate it. Maybe. But even if this does turn out to be the case, all is not lost. Let's see why.

MINDS, MACHINES AND EVOLUTION

On January 3, 1990, Tom Ray, a naturalist from the University of Delaware, pushed the start button on his computer, kicking into action a program called Tierra. After letting the program run all night, what Ray found in his machine the next morning was an electronic ecosystem of dazzling diversity—populations of many different types of organisms, all of which descended from a single ancestral organism that Ray had inserted into the program to get it started. As Ray stated, "From the most basic instructions there emerged an astonishing complexity." Such are the powers of evolution.

The Tierra simulator is an attempt to mimic some important aspects of Darwinian evolution in a machine. The organisms in this electronic ecosystem are self-reproducing strings of programming-language code. Each of these programs competes against the others for slices of the computer's processing time, as well as for memory locations in the machine, much as different organisms compete for an ecosystem's physical territory and food supply. So there is no a priori criterion imposed from the outside as to what is a "fit" organism. What's fit or unfit changes over time, depending upon how the organisms in the "soup" mutate, recombine and, in general, evolve so as to leave as many copies of themselves in the machine as possible. For details of the many clever ways Ray developed to ensure that a population of his "critters" captures characteristic features of a real ecosystem, the reader is referred to Ray's account of the whole experiment cited in the To Dig Deeper section.

The Tierra exercise was the first ever to demonstrate definitively that the process of evolution is independent of a particular material substrate. It can take place just as easily among a population of computer programs competing for memory space in a machine as it can among populations of carbon-based organisms competing to survive in an earthly environment. So why couldn't it happen in a population of machines?

Interestingly enough, this evolutionary argument seemed to be the position favored by Gödel himself, when asked whether his theorem posed an insurmountable barrier to the development of a true mechanical intelligence. The following reply seems to represent the sum total of Gödel's published words on the problem of strong AI:

> . . . it remains possible that there may exist (and even be empirically discoverable) a theorem-proving machine which in fact *is* equivalent in mathematical intuition [to the human mind], but cannot be *proved* to be so, nor even be proved to yield only *correct* theorems of finitary number theory.

By this remark, Gödel is suggesting that a machine equivalent in brainpower to the human mind might actually be created (e.g., by evolution). But if such a device did exist, Gödel's claim is that we would never understand it. It would be too complex for us.

So the Gödelian prescription is not to build a brain, but rather to *grow* one! And Ray's experiment shows that there's no logical barrier to following this dictum. What both Gödel and Ray are saying, in effect, is that a machine equal to a human in cognitive capacity will be just an example, albeit a very special one, of what we can call *artificial life* (A-life). Since this entire theme of "living machines" has become a hot topic of late, let me conclude this chapter by exploring a few of the striking parallels between the research agendas of the AIers and the A-lifers. Perhaps the best way to get started is by listing and commenting upon the following set of hypotheses underpinning belief in the existence of A-life, as put forth by Steen Rasmussen of the Los Alamos National Laboratory and the Santa Fe Institute, one of the principal players in the A-life game.

Postulate 1: *A universal Turing machine can simulate any physical process.* The content of this assumption is that the information-processing rules of any physical process can be mimicked by a suitably programmed computer. In short, the Turing-Church Thesis is true for physical systems.

Interestingly enough, this is exactly the assumption that Roger Penrose has called into question in his treatment of strong AI. An important part of Penrose's anti-AI argument is that the brain has ways of information-processing that transcend computability as characterized by a universal Turing machine. On the other hand, weakening this postulate to the statement that all human cognitive activity is computable yields the assumption that sustains the belief of pro-AI researchers in the ultimate success of their strong AI research programs.

Postulate 2: *Life is a physical process.* The crucial point here is the claim that life is a consequence of the functional organization of the different parts of a system, and that these functional aspects can be carried out in many different types of physical hardware. In particular, the relevant functional activities giving rise to life can comfortably be carried on within the confines of a computing machine.

Substituting the word *cognition* for the word *life* in the above statement leads to the functionalist position on strong AI. So if you're a believer in either Top Down or Bottom Up AI, there is nothing about thinking that transcends ordinary neurophysiological processes in the brain. Human thought then becomes a consequence of how the physical components of the brain are organized and does not involve the exact details of any particular type of hardware. Of course, connectionists hold that the wiring pattern linking the brain's neurons is important, but that that pattern can be duplicated in many distinct sorts of actual hardware—including a digital computer.

Postulate 3: *There are criteria by which we can distinguish between living and nonliving systems.* While all presently known conditions for life seem rather fuzzy, this postulate asserts that agreement can in principle be reached as to what is and isn't alive. In particular, all living systems should include the functional activities of metabolism, self-repair and replication.

In the AI context, this postulate gives rise to things like the Turing Test and Chinese Room arguments. How do we know someone is thinking? How do we know something is alive? In both cases there seem to be pretty strongly held intuitive ideas enabling us to say, in effect, "I know it when I see it." But the fur starts to fly when it comes to giving an explicit set of criteria that apply in all cases.

Postulate 4: *An artificial organism must perceive a reality R*, which for it is just as real as the "real" reality R is for us.* An important consequence of this assumption is

Postulate 5: *The realities* R* *and* R *have the same ontological status.* In other words, what we call reality is no more or less real to us than the reality seen by an artificial organism in a machine. So in a simulated rainstorm in the artificial reality, an artificial dog really does get wet.

Acceptance of these last two assumptions in either the AI or A-life worlds leads immediately to a plethora of issues surrounding "rights" for machines. If a genuine thinking machine has the same ontological status as a thinking human, then it's hard to see why we should not give such a device the same civil rights we accord to a human. Or so goes the argument, anyway.

Postulate 6: *We can learn about the fundamental properties of our reality R by studying the details of different R*s.* This means that looking at what artificial life does inside a machine can give us insight into what our human form of life is doing outside the computer.

This postulate is the raison d'etre for the entire A-life and AI undertakings. If we accept the machine version of neurons, thoughts, language or whatever as a valid representation of that same concept in *R*, it follows then that the machine-world version and the real-world version are isomorphic, or functionally equivalent, and whatever you learn from the study of one can be transferred to the other, mutatis mutandis. For example, this line of reasoning is what supports most of the interest in Tom Ray's Tierra simulator as a way of studying evolutionary processes.

So we come to the conclusion that if you want to study intelligence in a machine, it might be a smart move initially to shift attention from intelligence to life itself. If you can create life in a computer, then intelligence will almost assuredly follow.

As a final dollop of speculation, it's not hard to extrapolate the foregoing picture to its logical end in which machines evolve to the point where they are no longer interested in us lowly humans. Let's sketch one possible scenario for how this might happen.

Imagine setting a few thousand robots loose on the Moon, their prime directive being to mine, smelt, fabricate and assemble the materials necessary to build copies of themselves. These robots find the Moon a congenial place. Its low temperatures, abundant solar energy, lack of corrosive water vapor or gaseous oxygen and the abundance of silicon are just what the robots need to be fruitful and multiply.

If these robots are programmed to place high priority on self-reproduction, there will inevitably be competition for the raw materials and finished supplies. Such competition, in turn, leads to natural selection. In addition, we can ensure that the robot programs undergo regular mutation. For example, we could place an instruction in the ancestral program containing the imperative never to copy itself exactly. So each time a program is transferred, a substantial number of changes is made in it, changes that are determined randomly by, for example, waving a powerful magnet over the new robot's head or bombarding it with high-intensity X rays.

Soon the programs for these robots will become as incomprehensible to us as the mysterious statues on Easter Island or the indecipherable scribbles on the Voynich manuscript, and we can expect to find a large and autonomous robot civilization developing on the Moon. Who knows, perhaps some of the robots there will interest themselves in mathematics. At exactly this point, Gödel's theorem-proving machine, whose abilities equal the resources of human mathematical intuition, could come into existence.

On this perhaps unsettling note, we conclude our discussion of why forming our expectations of the ways and whys of the world by computation will always leave the door open to unanticipated events (i.e., surprises). It's just not possible to get it all by following a set of rules.

FIVE

THE IRREDUCIBLE

Intuition: Complicated systems can always be understood by breaking them down into simpler parts.

The whole is more than the sum of the parts.
—ARISTOTLE

Only wholeness leads to clarity,
And truth lies in the abyss.
—FRIEDRICH VON SCHILLER

Repetition is the only form of permanence that nature can achieve.
—GEORGE SANTAYANA

GETTING IT TOGETHER

Many readers of Mark Twain's story *Those Extraordinary Twins* probably thought he made up his account of the Siamese twins Luigi and Angelo, especially the part about Luigi being a heavy drinker and

Angelo a crusading teetotaler. So it may come as a surprise to know that Twain's tale is based on the lives of the first recorded real-life Siamese twins, Chang and Eng, who were born in Siam (where else?) in 1811. In fact, Chang really was a heavy drinker, while Eng was a teetotaler, a state of affairs that apparently caused more than a minor amount of family discord.

The story of Chang and Eng is an interesting one. They ended up as American citizens, taking the name Bunker, and before the Civil War they were slaveholders in North Carolina. Even more interesting is the fact that Chang and Eng had a passel of children: seven daughters and three sons for Chang, seven sons and five daughters for Eng. I'll leave speculation about the twin's methods in this regard to the reader's imagination.

The doings of Chang and Eng show that if you want to study the behavior of a system composed of many parts, breaking it apart into its component pieces and studying the pieces separately won't always help you in understanding the whole. I think it's clear that if Chang and Eng had been separated at birth as Siamese twins usually are today, Mark Twain would not have been the least bit interested; neither Chang nor Eng taken as an individual was especially noteworthy. What makes them special even today is the fact that they were two humans linked together in a very unusual fashion, a connection that led to interesting consequences. This connective structure led to a system considerably more complicated than that representing the typical man or woman. So in trying to understand the complicated system "Chang-and-Eng," it's essential to take this connectivity into account; Chang and Eng can't be understood by thinking of them as two disconnected individuals. A good summary of the general principle involved was given by the cybernetics pioneer W. Ross Ashby, when he remarked in 1956:

> Science today stands on something of a divide. For two centuries it has been exploring systems that are either intrinsically simple or that are capable of being analyzed into simple components. The fact that such a dogma as "vary the factors one at a time" could be accepted for a century, shows that scientists were largely concerned in investigating such systems as allowed this method; for this method is often fundamentally impossible in the complex systems.

It's exactly this kind of complexity and the surprise arising from connective patterns that we shall look into during the course of this chapter.

Geometry and Complexes

One way of thinking about connectivity is to consider a set of elements that may or may not share common properties. So, for example, we might have a set consisting of various types of flowers, along with a second set whose elements are different colors. The sharing of one or more colors would then establish a connective linkage between flower types. Alternately, a linkage between the colors could be made by focusing on those flowers that share a particular color. Let's get a bit more formal and see how to express these commonplace notions in the language of sets and relations.

Consider the set

$$X = \{\text{flowers}\} = \{\text{daffodil, rose, carnation, tulip, pansy, orchid}\}$$

$$= \{x_1, x_2, \ldots x_6\}$$

and the set

$$Y = \{\text{colors}\} = \{\text{red, yellow, green, blue, white}\} = \{y_1, y_2, \ldots y_5\}$$

As already discussed, a potentially interesting relation R for gardeners and florists might be

> "Flower type x_i is R-related to color y_j if and only if there exists a strain of flower x_i having color y_j."

Since normally there exist only yellow daffodils, we have only the pair (daffodil, yellow) in the relation R. A similar argument for roses shows that the pairs (rose, red), (rose, yellow) and (rose, white) are in the relation R. Continuing this process for all the other flowers in the set X, we can complete the entire relation R connecting the sets of flowers and colors.

A compact way of expressing the relation R is by means of what's called an *incidence matrix*, which we can represent by the symbol $\mathbf{R}$.

Let's label the rows of the rectangular array **R** by the elements of X and the columns by the elements of the set Y, agreeing that the element of **R** in the ith row and jth column equals 1 if the pair (x_i, y_j) satisfies the relation R; otherwise this element is 0. So for the flower example above, the incidence matrix **R** becomes

R	Red (y_1)	Yellow (y_2)	Green (y_3)	Blue (y_4)	White (y_5)
Daffodil (x_1)	0	1	0	0	0
Rose (x_2)	1	1	0	0	1
Carnation (x_3)	1	0	0	0	1
Tulip (x_4)	1	1	0	0	1
Pansy (x_5)	1	0	0	1	1
Orchid (x_6)	1	0	0	1	1

The incidence matrix is a way of representing the relation R that's particularly useful for calculating various measures of connection between the elements of the two sets under consideration. But when it comes to actually *seeing* the overall geometrical structure of the relation R, it's usually better to have a picture of the relation. Mathematicians have developed a standard procedure for constructing an abstract geometrical representation of any relation like R. This geometrical gadget is called a *simplicial complex*.

First of all, recall the relationship between a collection of abstract points and the geometric dimension of the space these points generate. A single point creates a zero-dimensional object; two points determine a line, which is a one-dimensional entity; three points form a triangle, the standard, or canonical, way of representing an area—a two-dimensional quantity. And, in general, $n + 1$ points fix an n-dimensional object for $n = 1, 2, \ldots$. Note that these standard geometrical objects—point, line, triangle, tetrahedron and so on—are just *abstract* constructs. They are used as canonical representatives for all objects of the same geometric dimension. Let's see how all these notions of dimension and complexes work in the case of the flowers-and-colors relation.

For notational compactness, we agree to label the elements of the set of flowers X by the symbols $x_1, x_2, \ldots x_6$, as was indicated earlier when we defined this set. Similarly, the colors are labeled $y_1, y_2, \ldots y_5$. Let's arbitrarily agree to call the set of colors the *vertex set*. From a geometrical point of view, we can think of each vertex as being simply a point in some abstract space.

Now consider the element x_1 (daffodil). Daffodils are only R-related to the color yellow, which is the vertex element labeled y_2. Therefore, x_1 is the 0-dimensional object (the point) consisting of the single vertex y_2. We draw this point on a piece of paper, labeling it y_2. Doing the same sort of analysis for the element x_2 (roses), we find that x_2 is the two-dimensional object formed by the vertices y_1, y_2 and y_5. Thus, rose is the *abstract* filled-in triangle formed by these three vertices. We also plot this triangle on our paper, making use of the previous vertex y_2, along with the two new vertices y_1 and y_5. Here we see a connection already appearing between the daffodils and the roses by means of their sharing the vertex y_2 (the color yellow). Continuing in this fashion, we generate an abstract geometrical object for each flower. These abstract points, lines, triangles and so on representing the various flowers are technically termed *simplices*, and the collection of all these simplices is called the *simplicial complex* of the relation R. This complex is an abstract picture of how the flowers are connected to each other by sharing colors. The geometry of this connective relationship is shown below in Figure 5.1. The above way of constructing the abstract geometry of a relation between two sets X and Y is entirely general, in the sense that it works for any relation R between any two finite sets X and Y. The key ingredient is to realize that the relation R "connects" each element in one of the sets with possibly several elements of the other. So if we agree to regard the elements of X as abstract points (i.e., vertices), then every element of Y acquires a dimensional character by being made up of a certain number of these points, the actual points being determined by the relation R. In

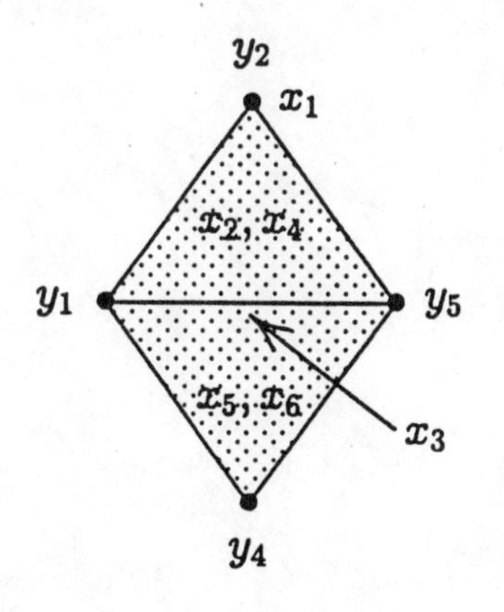

FIGURE 5.1. THE SIMPLICIAL COMPLEX FOR THE RELATION *R* LINKING THE FLOWERS AND COLORS

particular, if the element y in Y is R-related to $n + 1$ elements of X, then y is called an *n-simplex*, which has geometric dimension n. The general idea is shown schematically in Figure 5.2 for the case of a 3-simplex composed of the vertices A, B, C and D. Again, we call the reader's attention to the convention that it takes four points to determine a three-dimensional object (a volume), three points for a two-dimensional object (a triangle) having area and so on.

The fact that a simplex has a natural geometric dimension in which it "lives" suggests that trying to force anything associated with this object into a lower dimension is going to introduce some stress into the system. This point is illustrated by the familiar parlor game of trying to connect three adjacent houses to water, gas and electricity lines. This situation is shown in the top half of Figure 5.3. A bit of fumbling about soon leads to the conclusion that if the connecting lines all have to lie in a plane, it's impossible to make all the connections without the lines intersecting in at least one point like x. But as soon as we "lift" the problem up into its natural dimension three, we easily find the nonintersecting solution shown in the bottom half of the figure.

Since the idea of a simplicial complex and the connective structure it embodies is so central to our concerns in this chapter, let's look at another, somewhat less contrived, example to hammer home the underlying concepts.

Consider the seemingly interminable Middle East situation. Let's try to represent the conflict between the Arabs and the Israelis as a relation

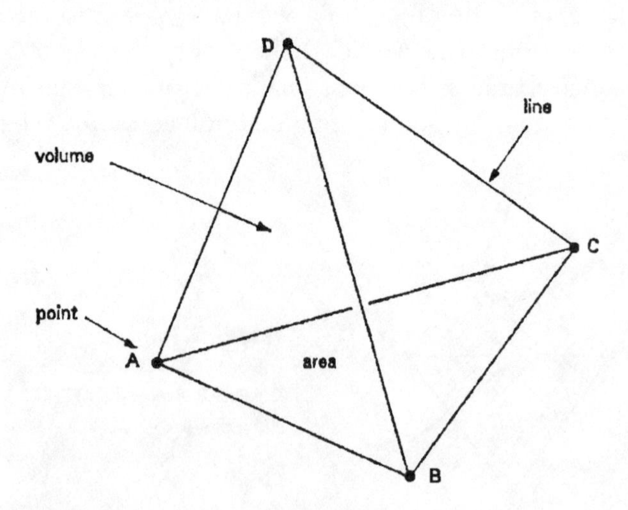

FIGURE 5.2. AN ABSTRACT 3-SIMPLEX

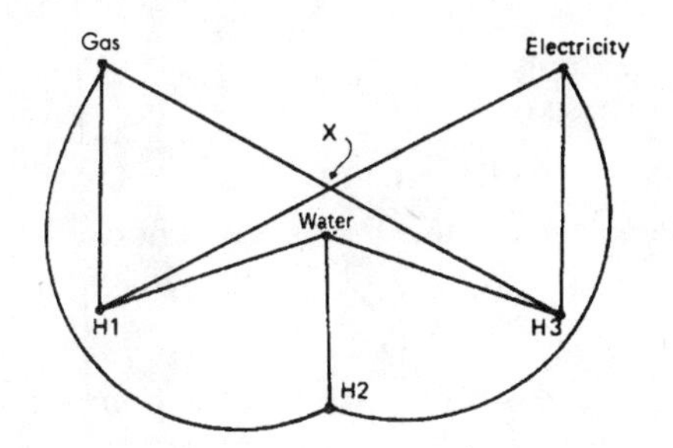

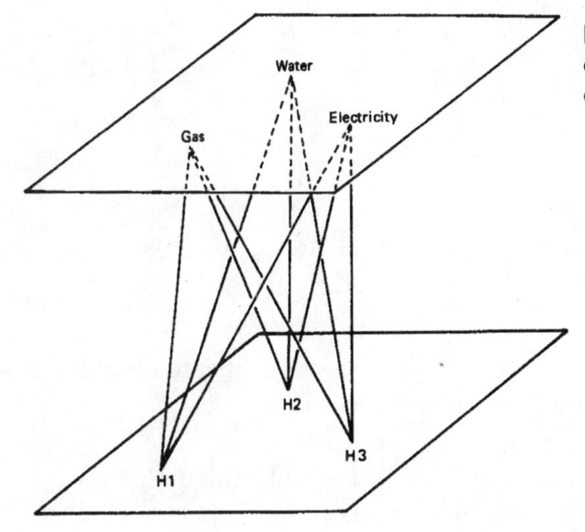

FIGURE 5.3.
GAS, WATER AND ELECTRICITY CONNECTIONS

between the set of participants and the issues that divide them. So for the set of issues X, we take

$$X = \{\text{issues}\} = \{x_1, x_2, \ldots x_{10}\}$$

where

x_1 = autonomous Palestinian state on the West Bank and Gaza
x_2 = return of the West Bank and Gaza to Arab rule
x_3 = Israeli military outposts along the Jordan River
x_4 = Israel retains East Jerusalem
x_5 = free access to all religious centers
x_6 = return of Sinai to Egypt

x_7 = dismantle Israeli Sinai settlements
x_8 = return of Golan Heights to Syria
x_9 = Israeli military outposts on the Golan Heights
x_{10} = Arab countries grant citizenship to Palestinians who choose to remain within their borders

For this example, we assume that the set of participants Y linked by these issues is

$$Y = \{\text{participants}\} = \{y_1, y_2, \ldots y_6\}$$

where

y_1 = Israel
y_2 = Egypt
y_3 = Palestinians
y_4 = Jordan
y_5 = Syria and Iraq
y_6 = Saudi Arabia

As a relation R characterizing the way the participants are connected to each other through the issues, let's use the rule

> "Participant y_i is R-related to issue x_j if and only if participant y_i is neutral or favorable toward issue (goal) x_j."

A possible incidence matrix for this relation is then

R	x_1	x_2	x_3	x_4	x_5	x_6	x_7	x_8	x_9	x_{10}
y_1	0	1	1	1	1	1	0	0	1	1
y_2	1	1	1	0	1	1	1	1	1	0
y_3	1	1	0	0	1	1	1	1	1	1
y_4	1	1	0	0	1	1	1	1	1	0
y_5	1	1	0	0	1	1	1	1	0	0
y_6	1	1	1	0	1	1	1	1	1	1

Examination of the complex whose vertices are the issues shows that the most likely negotiating partner for Israel is Saudi Arabia, which is neutral or favorable on all issues except one. This is because Saudi Arabia is the highest-dimensional object in this complex, being an eight-dimensional simplex consisting of all the issues except issue x_4 (Israel retaining East Jerusalem). However, both Egypt and the Palestinians are nearly as likely candidates since they are simplices of dimension only one less than Saudi Arabia. As the Camp David talks demonstrated some years ago, Egypt is indeed a favored negotiating partner due also to psychological and other factors not reflected in the above relation R.

While the foregoing ideas may look childishly simple, even primitive, we'll soon see that the notion of two sets and a relation between them contains a wealth of information about how the elements interrelate to each other via a connective pattern. But before moving on to a consideration of these matters, let's pause to take up a crucial point about the sets-and-relations setup that underpins this entire approach to structural modeling.

Everything we have said above rests upon the sets and relations being well defined. For instance, when we define a set of flowers we have to give a definite rule enabling us to determine whether or not a particular element belongs to the set. While this may seem an obvious requirement, defining a set in practice can pose considerable difficulties. For instance, if we want to define the set whose elements are the countries of the world, there is little difficulty in deciding that France, Canada and Brazil belong to the set, but what about places like Scotland or Bosnia-Herzogovina? It's easy to see how the matter of set membership can quickly turn into a very sticky affair.

Even after we've managed to put together some well-defined sets, there is still the problem of pinning down a meaningful relation between the elements of these sets. As we've already seen with the sets of flower types and colors, a relation in its most general form is simply a rule that assigns elements of one set to elements of another. But here again we run up against the problem of making the rule itself unambiguous (i.e., well defined).

To illustrate these considerations, suppose one of our sets is the members of a criminal gang, while the other set is a collection of criminal specialties, say things like counterfeiting, bank robbery, passing bad checks, jewel theft and extortion. A fairly obvious relation between these sets involves matching members of the gang to their criminal interests. Now comes the problem: Where does Louie, who works on bilking bereaved widows and lonely spinsters out of their worldly possessions, belong in such a categorization? His specialty seems to reside somewhere between the elements "jewel theft" and "extortion," with a slight bias toward the latter. So do we include both of the pairs (Louie, jewel theft) and (Louie, extortion) in the relation? Or do we just take one of them? Or neither?

These problems of making both set membership and set relations well defined are, of course, just a form of the far more basic issue of resolution and scale in the theory of measurement. It's clear that the problem of Louie and his low-life criminal interests would vanish into the now mythical luminiferous aether if we were to add another element to our set of specialties, the element "fraud." A similar refinement of scale would dissolve the difficulty of whether or not to regard Scotland as a country, by expanding what we mean by a country to include semiautonomous states that are part of the United Kingdom.

The point of this small digression is that the hardest part of many of the analyses we're going to look at in the following pages involves this initial step of defining sets and relations to make them relevant to the question at hand. Failing this step, we end up in a situation in which we literally don't know what we're talking about. With this caveat in mind, let's now get back to the business at hand, namely, looking for how surprises can emerge out of multidimensional connections.

Consider the primary color triangle shown in Figure 5.4, whose vertices are labeled Red, Blue and Green (the 0-simplices). The 1-simplices (Violet, Turquoise and Yellow), as well as the 2-simplex (White), are formed by combining the primary colors. So here we have a simplicial complex consisting of the single simplex $W = \langle R,B,G \rangle$, together with each of its faces. A person with normal color vision is capable of seeing any combination of the three primary colors at once, so will be able to see the entire spectrum of visible colors. Now let's do the following thought experiment: Take a subject who can see any combination of only two primary colors at a time. For the sake of definiteness, assume the subject cannot perceive any color containing red. Let the experiment consist of showing this subject flashcards, each of which is colored in one of the seven colors indicated in Figure 5.4. Assume that each of these seven colors appears randomly with equal likelihood. When a card is displayed, the subject tells us what color he sees, and we pass on to the next card. With a subject possessing

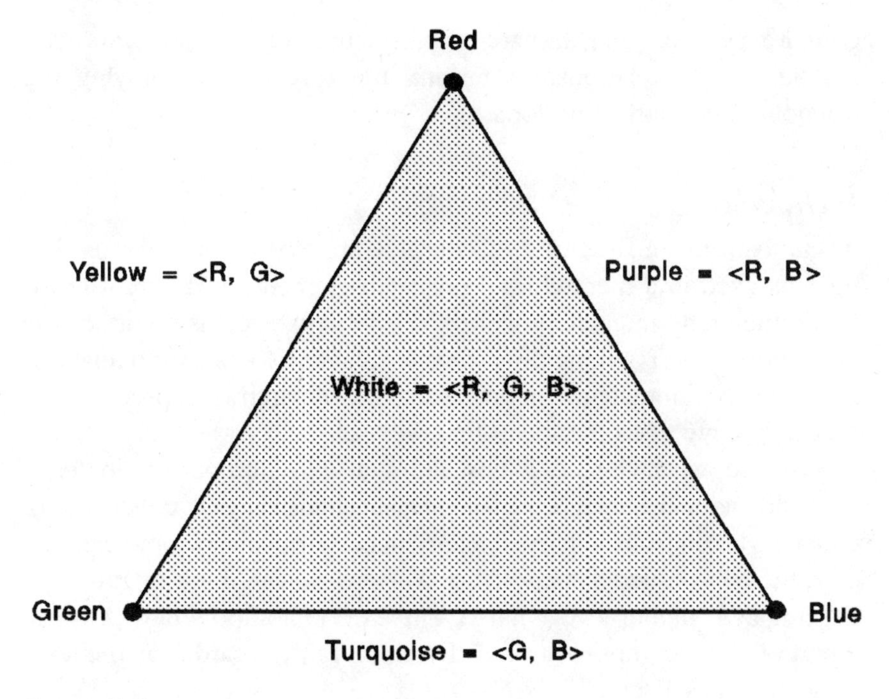

FIGURE 5.4. THE RGB COLOR TRIANGLE

normal color vision, we would expect each color to be identified correctly, and our records would then show that each color appeared one-seventh of the time, on the average. Now we ask what result would be expected from the subject who cannot see red.

For this subject, the set of possible responses is reduced from seven to three: Green, Turquoise and Blue. Therefore, whenever a flashcard is displayed that's one of the colors containing red, the subject will presumably name one of these three colors with equal likelihood. In short, the subject just guesses since he cannot actually see the correct color. So for such a color-blind subject, the records would show that G, T and B each appear one-third of the time rather than one-seventh, as was the case for a subject with normal color vision. Geometrically, this rearrangement of probabilities can be thought of as due to a reduction of the full color complex from its original two-dimensional form as the triangle $<R,G,B>$ to the one-dimensional complex $<G,B>$.

We'll see this same kind of phenomenon surfacing again when we consider geometrical aspects of classical probability theory later in the chapter. For now, let's look at another crucially important aspect of complex systems, their hierarchical structure, and how we can get a handle on the relationships among the levels by employing the foregoing geometrical notions.

Hierarchies

An almost universal feature of complex systems is that they tend to be organized in a hierarchical way, with elements at different levels in the hierarchy interacting to produce what we see as complication and complexity. Thinking of such systems in terms of sets and relations allows us to formulate the general idea of a hierarchy in a precise way. To do this, we need the idea of a *cover set.*

Suppose we have a set X containing a finite number of elements. Now consider a set Y, each of whose elements is a collection of elements taken from X. If every element of X appears in some element of Y, then Y is called a *cover set* for X. Here we can think of Y as a set existing at a "higher level" than X, since every component of X is just a part of some component of Y. If we arbitrarily regard N as the level of X, then Y is a set at level $N + 1$.

Since Y itself is a set, we can now consider a cover set Z for Y, together with an associated relation linking the two levels. We then regard Z as a level $N + 2$ set. This process can go in the opposite direction, too, letting X be a cover set for some set W at level $N - 1$. This idea of a hierarchy of sets and relations is shown abstractly in Figure 5.5 and more specifically for the case of a hospital in Figure 5.6.

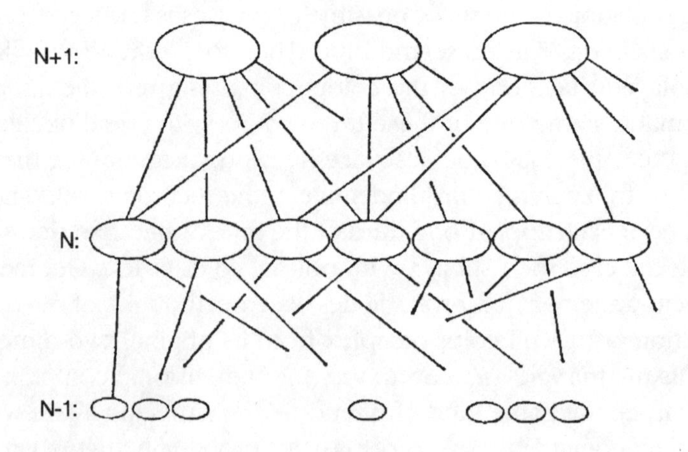

FIGURE 5.5. A HIERARCHY BASED ON COVER SETS

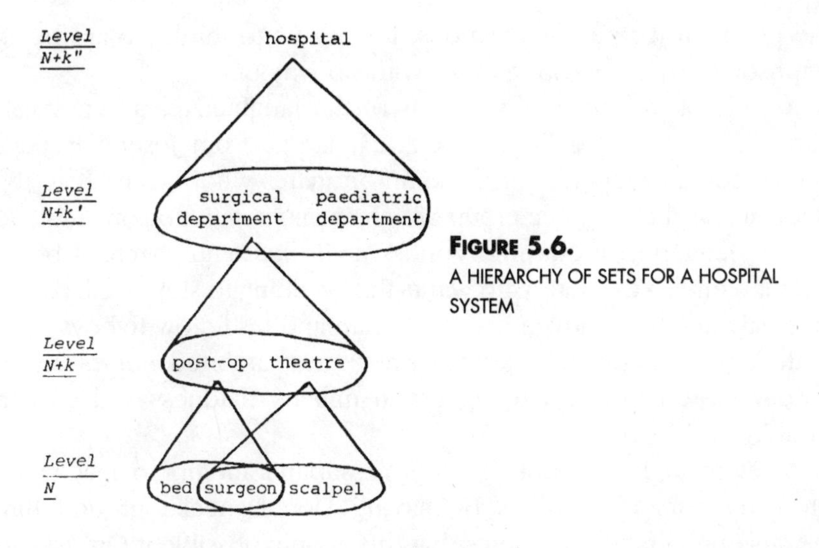

FIGURE 5.6.
A HIERARCHY OF SETS FOR A HOSPITAL SYSTEM

So, for example, in the hospital case we have the single-element set {hospital} existing at level $N + 3$. That set in turn acts as a cover for the sets of departments at level $N + 2$, as well as for the level N set consisting of the elements "bed," "surgeon" and "scalpel."

Basic as the above ideas are, they already contain enough mathematical meat for us to say some interesting things about matters in the arts and humanities, areas often (erroneously) thought to be beyond the boundaries of mathematical investigation. Let's look at a couple of examples.

In his book *Multidimensional Man*, Ron Atkin suggests that the process of evoking laughter or tears from a particular N-level situation corresponds to a movement either up to level $N + 1$ for laughter or down to level $N - 1$ for tears. His argument is that in order to be aware of witticisms present at level N, we must be able to contemplate new relationships on the N-level sets, either by rearranging existing elements or by extending the elements to find new relationships between them—in short, by being aware of level $N + 1$. Atkin's claim is that it's this sudden jump to the level $N + 1$ set that generates a release of laughter.

In contrast to laughter, which is a widening of our horizons by a movement *up* the conceptual ladder, sorrow and tears represent a movement *downward* that shrinks those horizons. Moving up the hierarchy, we see the possibility for new relationships, a potentially liberating situation. A movement downward contracts or eliminates the potential for new interactions, forcing us to think we are being imprisoned by the existing order with no way out.

As one of the many illustrations of the laughter/tears hypothesis cited in Atkin's book, let's look at a passage from Joseph Heller's classic novel *Catch-22*. For those unfortunates who haven't read the book, it tells the story of a group of fighter pilots stationed on an island in the Mediterranean during World War II, the main character being a pilot named Yossarian who wants the bombing to stop. Heller's tale is mostly about the adventures Yossarian and his fellow flyboys have with American generals, Roman prostitutes, assorted nurses and a motley crew of other participants in military madness and civilian insanity.

A key scene in the book involves Yossarian's attempt to get Orr out of flying combat missions by having Doc Daneeka ground him because he's crazy. Doc states that this is only possible if Orr asks to be grounded. But Doc then adds that he will not be able to ground Orr if Orr asks to be grounded, since the request itself will constitute evidence that Orr isn't crazy.

For our set-theoretic purposes, let's think of Heller's passage as being an N-level situation, involving the *individuals* Yossarian and Orr. At level $N+1$ we have a set consisting of a number of descriptive words like Sane, Missions, Grounded, and Fit for Duty. Finally, we find that at level $N+2$ there is a set consisting of the single element Doc, since this is the agent who can decide whether or not a man at level N is a member of the $N+1$-level element Fit for Duty.

The scene's humor comes from the fact that Yossarian thinks he's "covered" by the words Insane and Flying Missions at level $N+1$, which would automatically mean that he cannot also be covered by the term Fit for Duty. But Doc reorganizes the cover set at level $N+1$ by saying that Yossarian's request is by itself sufficient to demonstrate that he's Sane, therefore covered by Fit for Duty. Here we see Yossarian's frustration at feeling trapped inside the N-level set and having his appeal to the $N+2$-level set rejected through a rearrangement of the $N+1$-level cover. So if you identify with Yossarian, you're brought to the verge of tears. But if you stand outside the book—at, say, level $N+3$—then you experience the urge to laugh at this "Catch-22" situation.

To close this short discussion of hierarchical structure, let's look at how thinking hierarchically can help unravel logical puzzles of another type by revisiting a version of the famous Epimenides Paradox that played such an important part in our story of Gödel's Theorem in Chapter Four.

The famous Barber Paradox involves a village in which the town barber shaves all those men who do not shave themselves. Since the barber himself is a man, it seems to make sense to ask if the barber shaves himself. Tracing through the logical possibilities, we come to the surprising conclusion that the barber shaves himself if and only if he *doesn't* shave himself. Considering this situation hierarchically allows us to dissolve the paradox.

As our basic sets we take B = {barber} = $\{b\}$, a set consisting of the single element "barber." We also have the set $M = \{m_1, m_2, \ldots m_k\}$ representing the men in the village. The obvious relation R is given by the rule

> "(b,m_i) is in the relation R if and only if the barber shaves man m_i."

So why can't we determine whether or not the element (b,b) is in the relation R ?

The problem here is to recognize that the barber is only a barber (i.e., is properly defined) insofar as he shaves people. Thus, the element "barber" of the set M is really a symbol for a subset of men—namely, the set of men who are shaved by the barber. Thus, the barber *as a barber* really exists at the level $N + 1$, say. And when we try to put him in among the set M, we fail because we are trying to regard an N + 1-level object as an N-level object, which is what leads to the paradox.

Hierarchical analyses based upon the notion of cover sets, interesting as they appear, are limited in what they can tell us about the overall manner in which a relation R binds together the elements of two sets. For more information, we need to develop a way to measure the global connective pattern among the objects in a complex. This leads to the idea of what's termed *q-connectivity*.

MAKING CONNECTIONS

In his 1977 book *The Luck Factor,* author Max Gunther identifies five factors that he claims separate the lucky among us from the less fortunate. Gunther terms one of these factors the Spider Web Structure, which amounts to what in today's terms we would probably call networking. In other words, if you want to be lucky it helps to create a dense web of friendships and contacts that you can plug into for getting "lucky vibes." Whether or not belonging to a network of connections makes you "lucky" is a matter of semantics, I suppose, but it's hard to argue against the underlying principle that the odds of Dame Fortune smiling your way are better if you know a lot of people than if you don't. This principle also serves to illustrate the notion of chains of connection linking simplices in a complex. Just as in a spider web, where two distant parts may mutually resonate with each other without being in direct physical contact, in a simplicial complex we may have two parts affecting each other at a given dimensional level without the two simplices having even a single vertex in common. In this section we see how this can happen and what it means for the possibility of surprising behavior to emerge from the complex.

The simplices of a complex are connected to each other by sharing vertices. But this does not mean that any pair of simplices have vertices in common (i.e., that the simplices are connected *pairwise*). It's perfectly possible for two simplices to have no vertices in common and yet to be connected to each other by an intermediate chain of simplices that serves as a bridge between them. Figure 5.7 shows the general idea.

With this picture in mind, we say that two simplices are *q-connected* if the lowest-dimensional object in any chain linking them has dimension q. In other words, two simplices are q-connected if q is the weakest link, dimensionally speaking, in any chain linking the two

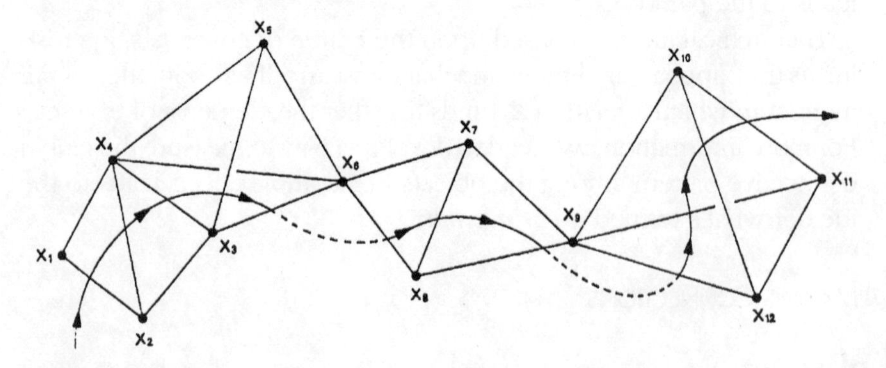

FIGURE 5.7. CONNECTIVE STRUCTURE IN A COMPLEX

simplices. Let's illustrate this crucial idea using the flowers-and-colors example considered earlier.

Consider the complex each of whose simplices is one of the types of flowers. We've already seen that these objects are connected to each other by sharing vertices, each of which represents one of the possible colors. Let's first consider matters at dimensional level $q = 2$. This means that we're looking for pairs of flower types connected to each other by a chain whose weakest link has geometric dimension two (i.e., consists of at least three vertices). Again let me emphasize that two flowers might be in the same 2-chain *even if they have no colors in common*. All that's required is that there be a sequence of flowers such that adjacent members of this sequence share at least two colors and that our target flowers form the first and last elements of the sequence.

Unfortunately, the flowers-and-colors complex is too simple to reveal this possibility. A quick glance at the incidence matrix shows that the pair (rose, tulip) and the pair (pansy, orchid) are 2-connected by sharing three colors. But in this case the two flowers involved are directly connected at level $q = 2$, thus forming a chain having only two links (i.e., there are no intermediate flowers). Similarly, it's not hard to see that the complex has only a single component at dimension level $q = 1$, consisting of all the flowers except daffodil. Finally, there is again only a single component at level $q = 0$, but this time it consists of all six flower types. This implies that for every flower there is a 0-chain connecting it to every other flower. So all the flowers are connected to each other by sharing at least one color.

While the flowers-and-colors example gives a hint of what we can hope to learn about the overall connective structure of a complex by employing this kind of *q-analysis*, the sets of flower types and colors have too few elements for any rich pattern of connections to emerge. So let's turn our attention to some more elaborate examples that allow us to exploit the ideas of global connectivity in interesting everyday settings.

Ecosystem Food Webs

Since species survival and extinction is one of the most basic concepts in ecology, it's natural for ecologists to place great emphasis in their work on what are called *predator-prey networks*. Figure 5.8 shows the

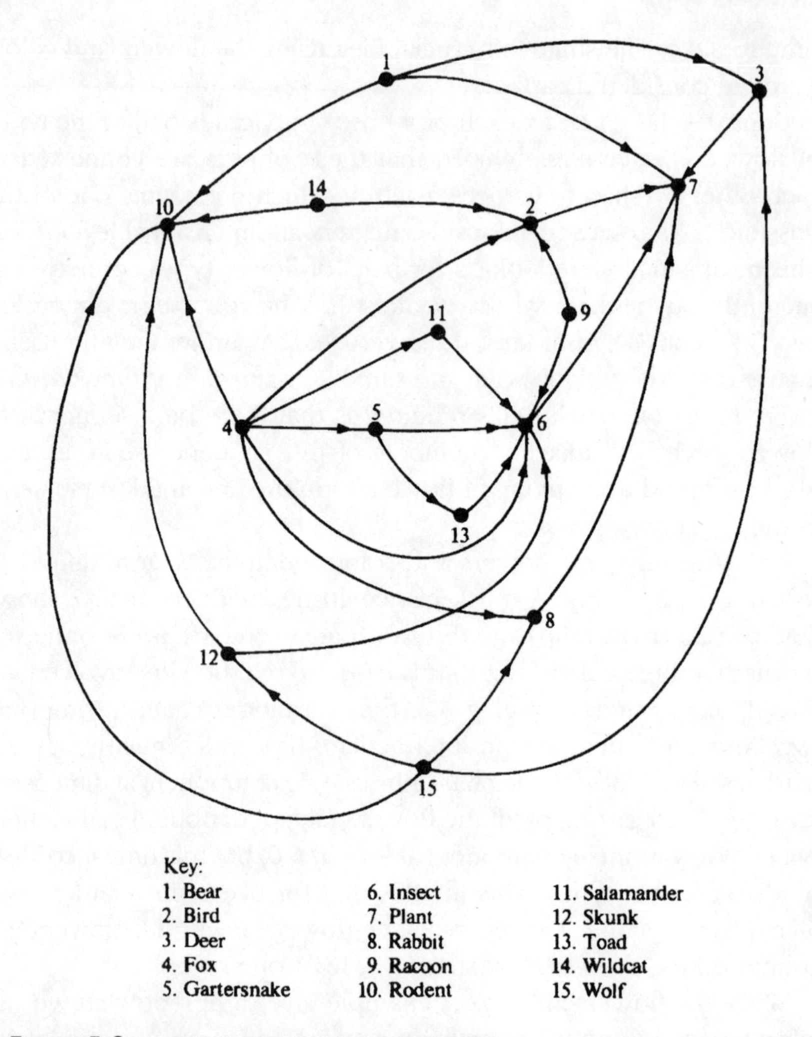

FIGURE 5.8. GRAPH OF A PREDATOR-PREY NETWORK

graph of a hypothetical network of this type containing fifteen species of plants and animals. In the graph, there is an arc directed from species i to species j if i feeds upon j.

The problem with this graphical depiction is that it doesn't show any of the multidimensional structure of the predator-prey interaction. So we'll use the techniques of q-analysis to tease out some of these higher-level connective patterns.

First, we need some sets and a relation. Letting the fifteen species be represented by the set $X = \{x_1, x_2, \ldots x_{15}\}$, we can define a predator relation λ_{PRD} by the rule

"(x_i, x_j) is in the relation λ_{PRD} if and only if x_i is a predator of species x_j."

Here it's of interest to note that the relation λ_{PRD} is defined using the single set X twice. In other words, to apply our methods, the two sets involved need not necessarily be different sets.

We saw earlier that for purposes of computation it's usually easier to represent a relation algebraically as an incidence matrix rather than listing pairs of elements or trying to draw its geometrical form (which cannot be shown anyway unless the complex is of dimension less than four). So here is one plausible incidence matrix for the relation λ_{PRD}:

λ_{PRD}	x_1	x_2	x_3	x_4	x_5	x_6	x_7	x_8	x_9	x_{10}	x_{11}	x_{12}	x_{13}	x_{14}	x_{15}
x_1	0	0	1	0	0	0	1	0	0	1	0	0	0	0	0
x_2	0	0	0	0	0	0	1	0	0	0	0	0	0	0	0
x_3	0	0	0	0	0	0	1	0	0	0	0	0	0	0	0
x_4	0	1	0	0	1	1	0	1	0	1	1	0	0	0	0
x_5	0	0	0	0	0	1	0	0	0	0	0	0	1	0	0
x_6	0	0	0	0	0	0	1	0	0	0	0	0	0	0	0
x_7	0	0	0	0	0	0	0	0	0	0	0	0	0	0	0
x_8	0	0	0	0	0	0	1	0	0	0	0	0	0	0	0
x_9	0	1	0	0	0	1	0	0	0	0	0	0	0	0	0
x_{10}	0	0	0	0	0	0	0	0	0	0	0	0	0	0	0
x_{11}	0	0	0	0	0	1	0	0	0	0	0	0	0	0	0
x_{12}	0	0	0	0	0	1	0	0	0	1	0	0	0	0	0
x_{13}	0	0	0	0	0	1	0	0	0	0	0	0	0	0	0
x_{14}	0	1	0	0	0	0	0	0	0	1	0	0	0	0	0
x_{15}	0	0	1	0	0	0	0	1	0	1	0	1	0	0	0

A bit of tracing though the connections in this incidence matrix turns up the geometrical fact that the simplicial complex whose simplices are the fifteen species has the following number of *disconnected* components at various dimension levels q:

$q = 5$; one component
$q = 4$; one component
$q = 3$; two components
$q = 2$; three components
$q = 1$; two components
$q = 0$; one component

So what can we say about the structure of this food web on the basis of these numbers?

Consider the lowest-dimensional level $q = 0$. At this level, the network has a single connected component. Recalling the definition of the relation λ_{PRD}, this means simply that there are no isolated species in the ecosystem; everyone is connected to everyone else by at least one predator. Moving up to dimensional level $q = 1$, there are two components. We can interpret this as saying that the ecosystem splits into two disconnected components when we consider those species connected to each other by a chain in which each element shares at least *two* predators. Similar conclusions follow for the other dimensional levels.

Analysis of this type gives us information about the geometry of the network taken as a whole. This is sometimes termed the system's *global* geometry. But it tells us very little about how well each individual species is integrated into the overall ecosystem. For this sort of *local* information, we need another kind of measure, sometimes called the *eccentricity* of a simplex. The basic idea is rather straightforward.

Consider a single species in the food web, represented by a simplex x. Intuitively, we would say that x is not well integrated into the overall network if it has many vertices that are not part of any other simplex in the complex (i.e., if x does not share many prey with other species in the system). So if dimension $x = n$ (i.e., x is comprised of $n + 1$ vertices), and if m is the largest number of vertices that x shares with any other single simplex in the complex, then the difference $n - m$ is a measure of how eccentric x is as a member of the ecosystem. However, it's also reasonable to assume that this difference is more significant at lower-dimensional levels than at higher ones since the relative difference is what really matters. So we normalize the difference $n - m$ by making it relative to m. This leads to our final measure of eccentricity as

$$\text{ecc}\,(y) = \frac{n - m}{m + 1}$$

where the number 1 appears in the denominator to avoid possible zero divisors that arise for those species x (like plants, vertex x_7) that share *no* prey with other species in the system.

Using this formula, we can compute the eccentricity of each species in our food web. These numbers turn out to be:

Species	x_1	x_2	x_3	x_4	x_5	x_6	x_7	x_8	x_9	x_{10}	x_{11}	x_{12}	x_{13}	x_{14}	x_{15}
Eccentricity	$\frac{1}{2}$	0	0	2	1	0	∞	0	0	∞	0	0	0	0	1

These values bear out many of our intuitions about this network, such as that species like plants (x_7) and rodents (x_{10}) that do not prey on any of the species are very eccentric indeed—at least from the perspective of being a predator. Of course, if we had defined our original relation in terms of prey rather than predation, then we would have seen a quite different picture of the situation. But I'll leave this exercise to the reader. For now, let's shift our attention from the forests and savannas to the roar of the grease paint, showing how the same brand of mathematical medicine can be used to shed light on the structure of a Shakespearean play.

A Midsummer Night's Dream

Even the most casual theater-goer knows that a play has a basic structure apart from its obvious division into acts and scenes. This underlying structure involves the twists and turns of the plot, carried out by the development of characters and situations. It's this structure that both the actors and the audience intuitively recognize when they call a play "good." Here we use ideas of connectivity to study this "goodness" in the context of Shakespeare's comedy *A Midsummer Night's Dream*.

Based on one edition of *A Midsummer Night's Dream*, the play can be divided into three sets:

A = {the play, acts, scenes and subscenes}
B = {the characters}
C = {the commentary, plots, subplots and speeches}

The components of the above sets induce a hierarchical ordering on the play. So, for example, taking set A we have the play itself, existing at level $N + 2$. Then there are the acts, which live at level $N + 1$. In the version of the play I'm reading, the editor has divided it into five such acts. (Note: Shakespeare himself did not necessarily divide his plays into acts and scenes; these categories were adopted by later editors.) Moving further down the hierarchy, at level N there are nine scenes, while at level $N - 1$ we have twenty-six subscenes based on major changes in the composition of the play and the entrances and exits of characters.

To illustrate the connective structure of *A Midsummer Night's Dream*, let's focus our attention on just one of the many relations existing among the foregoing sets: the N-level relation between the play's characters and the scenes. Suppose we let X denote the set of characters, of which the play contains twenty-one (Hippolyte, Theseus, Hermia, Puck et al). We then let Y represent the set of nine scenes. The obvious relation is that a scene is related to a character if that character appears in the given scene.

Now we have to choose whether we want to let the scenes be the simplices of our complex, each of which are formed by vertices consisting of characters, or have things the other way around. Let's agree to look at the play from the point of view of the scenes, taking the set of characters X to be the vertex set while letting the scenes form the simplices. Under this assumption, q-analysis can uncover some of the global geometrical structure in the relationship between the scenes and characters. Carrying out such an analysis shows that the play is dominated by the final scene, which sees twenty of the twenty-one characters making an appearance. This scene, of course, is the culmination of Theseus and Hippolyte's wedding celebrations in the performance of the play by Bottom and his friends. Scene 7 is the next-highest-dimensional object, involving the appearance of fifteen characters. This is the scene dealing with the outcome of Oberon's trick on Titania and their reconciliation. These two scenes are the most eccentric, as well. It turns out that the complex whose simplices are the scenes only becomes integrated into a single component at dimensional level $q = 5$. This means that an audience has to follow six characters in order to be aware of all nine scenes.

In a production of *A Midsummer Night's Dream* directed by Peter Brook of the Royal Shakespeare Company, two characters, Theseus and Oberon, were combined to form one character. This production also combined Hippolyte and Titania into a single character. Since in Brook's production these characters were played by a single actor and actress, respectively, the question arises as to whether this affected the way the audience saw the play. Our geometrical ideas offer one way of addressing this question.

Suppose we define the new set of characters X, which now consists only of the nineteen characters in the Royal Shakespeare Company's production, keeping the set of scenes as before. Again carrying out the q-analysis, we find no significant change. Thus, an audience focusing attention on the scenes would see the play as being more or less the same in either the conventional or the "slimmed down" productions. But if we shift attention to the characters, taking the set of scenes Y as the vertices and the elements of X as our simplices, the situation changes considerably.

In the standard case of twenty-one characters, the structure of the play is dominated by two components at dimension level $q = 5$. So an audience following six scenes sees the play as essentially about Demetrius and Helena on the one hand and about Puck on the other. At dimension level $q = 3$, the characters fall into two components again. So a theater-goer following four scenes sees the play as concerning either the lovers Puck, Oberon, Titania and the fairies or as about Bottom and his friends. But what about Peter Brook's "avant garde" version of the play?

In the production with only nineteen characters, the play's structure is dominated by three components at dimensional level $q = 5$. So a viewer following six scenes sees Brook's production as being about Theseus/Oberon, Demetrius and Helena or as about Hippolyte/Titania or as about Puck. So Peter Brook's combining of roles has increased the importance of the Hippolyte/Titania character, which was previously in the structure only at level $q = 4$. This production enhances the Theseus/Oberon character as well, which also appeared in the earlier structure only at level $q = 4$.

There is much more that can be said through this kind of geometric analysis of *A Midsummer Night's Dream*, but space constrains us. So let's reluctantly move now from the theater to the world of coffee houses and chessboards to see how our geometrical ideas of connectivity enter into one of mankind's oldest pursuits—capturing the King.

The Geometry of Chess

Emmanuel Lasker reigned as the world chess champion for twenty-seven years. Unlike many chess geniuses, Lasker's interests were far from narrow, and his concern with philosophical matters led to a deep consideration of what he called the "philosophy of struggle." For Lasker, the chessboard was a stage reflecting the struggle of life in its purest form, a view encapsulated in his well-known remark, "On the chessboard lies and hypocrisy do not long survive." He went on to note that "there are sixty-four squares on the chessboard; if you control thirty-three of them you must have an advantage." While this is a vast oversimplification of the situation, it points out the importance of positional play in the thinking of chess masters. This positional, or strategic, view of the game also suggests that it should be possible to use the multidimensional perspective of *q*-analysis to evaluate board positions. Let's see how this might be done.

The game of chess can be considered as a relation between the squares of the board and the Black and White pieces. Actually, there are at least two important relations here: (1) a relation R_B between the Black pieces and the squares and (2) a relation R_W linking the White pieces and the squares. But these relations might take many forms. For instance, we could define a relation by saying that a White piece and a particular square are related if that piece occupies the given square. However, such a relation, while perfectly well defined, is quite useless for the simple reason that it doesn't embody any of the rules of the game of chess.

In order to incorporate the rules of the game into some meaningful relations R_W and R_B, let's first define what we mean by a man (pawn, Knight, Bishop, Rook, Queen or King) *attacking* a given square. (Note: Here we shall adopt standard chess jargon, calling all the chessmen *men* while reserving the term *pieces* for those men that are not pawns.) For the sake of definiteness, let's center attention for the moment on the White men. We say that the White man P *attacks* square S if exactly *one* of the following conditions holds:

1. If it's White's move and P is not a pawn or the White King, then $P \rightarrow S$ is a legal move.
2. If P is a pawn, then S is a capturing square for P (this means that S is one of the two squares diagonally in front of P).
3. If there is a White man Y on square S, then P is protecting Y (in the ordinary chess-playing sense of one man protecting another).
4. If P is the White King, then S is adjacent (horizontally, vertically or diagonally) to the square occupied by P.
5. If it's White's move and square S contains a Black man Z (other

than the King), then P captures Z is a legal move. (Note: If Z is the Black King, we are at checkmate and the game is over.)

6. The Black King is on square S and is in check to P.

With this idea of *attacking* in mind, we then define the relation R_W linking the White men and the squares of the board as follows:

> "White man P_i is R_W-related to square S_j if and only if P_i is *attacking* square S_j."

The corresponding relation R_B for the Black men is defined analogously. Note that in these relations a piece does not attack its own square. This means that the pieces and pawns must protect each other by attacking the occupied squares. Now let's see how these relations can be used to understand the relative strengths of the pieces and to examine how they combine their varied abilities to form a "field of force" at any particular stage of play.

Let's agree to call a particular distribution of the White and Black men on the board a *mode*. The game begins with White to move in mode [0, 0]. After White makes the first move, the game enters mode [1, 0]; after Black's first move the game is in mode [1, 1] and so on. Figure 5.9 shows the distribution of the playing pieces when the game is to begin in mode [0, 0]. Let's consider the relation R_W, expressing the way the White men are linked by attacking squares on the board.

In this starting position, the QR-pawn on square a2 is a 0-simplex, since it attacks the single square b3. Similarly, the KR-pawn is also a 0-simplex. The remaining White pawns are all 1-simplices (e.g., the Q-pawn is the 1-simplex attacking squares c3 and e3). The White Queen, on the other hand, is a 4-simplex attacking squares c1, c2, d2,

FIGURE 5.9.
MODE [0, 0] AT THE BEGINNING OF A CHESS GAME, WHITE TO MOVE

e2 and e1. In this [0, 0] mode, the Q and K are 1-connected by both attacking squares d2 and e2. Carrying out this connectivity analysis for dimensional levels q less than or equal to four, we find that in mode [0, 0] the q-connected components of the complex formed by the White men are:

$q = 4$; two components: Q and K
$q = 3$; two components: Q and K
$q = 2$; four components: Q, K, QN and KN
$q = 1$; eleven components: {KN, KNP}, {QN, QNP}, {Q}, {K}, {QR}, {KB}, {QB}, {QBP}, {QP}, {KP}, {KBP}
$q = 0$; three components: {KR}, {QR}, {remaining White men}

Dimensional analysis of this sort yields a new perspective on the traditional arguments for the relative strengths of the playing pieces. Taking the value of a pawn to be P = 1, the traditional values assigned to the pieces are B = N = 3, R = 5, Q = 9, K = infinite. Leaving the King out of consideration, we can reasonably measure the relative strengths of the pieces in any playing mode by their dimensions. In mode [0, 0], these values are Q = 4, KN = QN = 2, with all the rest of the men having dimension 1 except for the KRP and the QRP, which have dimension 0. Of course these values shift during the course of the game. For example, if White makes the King's pawn opening move e2 → e4, the game enters mode [1, 0], and it's easy to see that the dimensions of all the White men remain the same except for the Q and KB. In this new mode, the dimension of the Queen is now 7 instead of 4, while the KB has increased its dimension from 1 to 5.

Following the above line of argument, it's instructive to consider what the *maximum* possible dimensions of the pieces can be in the absence of obstacles on the board. R. H. Atkin has computed these quantities to be as follows:

Piece	***Maximum Dimension***	***Maximum Attained***
Pawn	1	in all modes
Knight	7	when N is inside the square c6-f6-f3-c3
Bishop	12	when B is on square d5, e5, d4 or e4
Rook	13	when R is on any square
Queen	26	when Q is on square d5, e5, d4 or e4

Comparing these maximum dimensions with the traditional values given earlier, we find pretty good agreement, with the exception of the greatly increased value of the Bishop as compared with the Knight. We notice also the great importance of the central squares d5, e5, d4 and e4. Various theories of chess openings have emphasized the significance of controlling these squares. The dimensional analysis also enables us to appreciate schools of thought devoted to the argument that opening play should be dedicated to the task of bringing the Rook into play as soon as possible. Castling, as well as the classical King's Gambit opening with its early sacrifice of the KBP, are devices for accomplishing just this. This kind of analysis enables us to attach some geometrical structure to any particular mode of the game. Let's see briefly how this goes by considering a specific game, what in the chess world is called the *Immortal Game*.

Figure 5.10 shows the board position after the twelfth round of play in the famous Anderssen-Kieseritsky game played in London in 1851. Consider first White's (Anderssen's) view of the board. This involves the complex whose vertices are the squares and whose simplices are the White pieces. In this complex, we have dim (where dim stands for dimension) White Queen = 8, dim KN = 7, dim K = dim KR = 4 and dim QB = 3. Moreover, the White pieces are not totally connected until we reach dimensional level $q = 1$, and even then the only pieces that are connected are the Q, K and KR. We also find that this complex consists of five separate components at the lowest level $q = 0$.

Black's view of this board leads to a complex very similar to that of White. But things look quite different when we shift attention to the *conjugate relations*, which are defined by looking at the board's view of the playing pieces. In this case, squares that are deep inside the Black camp and close to the Black K are already part of White's structure—either as simplices in White's view of the board or as

FIGURE 5.10.
MODE [12, 12] IN THE ANDERSSEN-KIESERITSKY GAME

vertices in the board's view of White (the conjugate complex). Moreover, the Black Q is under attack by the White pawns, a very dangerous situation for Black. So, in effect, the positions of the King-side pawns and the KN causes the Black Queen to act as if it were the 3-simplex consisting of the squares <h6, h5, g5, g4>. But each of these vertices is in the board's view of White. In other words, the Black Queen cannot move without being captured.

Regrettably, there's no space here to enter into more details of the way the above sort of analysis can be employed to assess the strategic value of the various pieces at any stage of play. Nor can we consider further how this information might be employed in developing a tactical procedure for deciding what moves to make. The chess aficionado yearning for more detail will find ample consideration of these matters in the material cited in the To Dig Deeper section for this chapter. The main point here is to emphasize the way the connectivities linking the pieces and the squares change during the course of play and how this ever-shifting mosaic takes place in a high-dimensional space difficult to envision in terms familiar from our everyday experiences of ordinary space and time. And speaking of time, the time has come to turn our attention to how this same circle of geometrical notions allows us to look at time itself as a multidimensional phenomenon.

THE TIME OF YOUR LIFE

By now, just about everyone knows that on April 15, 1912, the *Titanic* sank on her maiden voyage, resulting in the loss of over 1500 of her 2207 passengers and crew. What is perhaps not so well known is that the *Titanic*'s tragic fate appears to have been foreseen in various dreams, hunches, trancelike visions and even in two novels.

One of the most intriguing precognitive images of the *Titanic* disaster occurred on April 10, 1912, the day the *Titanic* left Southampton for her journey to New York. On that day, Mrs. Jack Marshall was standing on the roof of her house, watching the ship's passage through the narrow body of water separating England from the Isle of Wight. Mrs. Marshall suddenly turned to her family, who were watching with her, and stated in a very agitated voice:

> That ship is going to sink before it reaches America . . . Don't stand there staring at me! Do something! You fools, I can see hundreds of people struggling in icy water! Are you all so blind that you are going to let them drown?

Despite assurances from her family that the *Titanic* was unsinkable, Mrs. Marshall remained in an agitated state until five days later when her deadly vision was fulfilled.

What is especially striking about the numerous "Mrs. Marshalls" who foretold the *Titanic*'s grisly end is the apparent elasticity of time. Somehow time seems to reverse its normal flow in such accounts, enabling otherwise average people to get a glimpse of the future.

Even more familiar to most of us is the feeling that time has a qualitative character about it. While the Newtonian view of time as the regular oscillation of some giant pendulum in the sky seems to work well in physics, everyday life is filled with expressions like "time just got away from me today" or "time's hanging heavy on my hands," statements suggesting a far more subjective aspect to our perception of time than Newton ever conceived of. This idea is well summarized in the verse by Guy Pentreath:

> For when I was a babe and wept and slept, Time crept;
> When I was a boy and laughed and talked, Time walked;
> Then when the years saw me a man, Time ran,
> But as I older grew, Time flew.

In the next few pages we'll explore the way important aspects of Pentreath's poetic vision of the elasticity of time can be captured by thinking of temporal duration as intervals between the occurrences of multidimensional events.

The idea that ordinary space is composed of relations between objects is an old one, dating back at least as far as Aristotle and championed later on by the German polymath Gottfried Wilhelm Leibniz, among others. The alternative view, in which space is given a priori and objects are "things" that merely sit in it, rose to popularity with Descartes and forms the basis for Newtonian science. It seems inevitable that this Newtonian view of space should be accompanied by a similarly absolutist Newtonian vision of time, as is indeed the case. In this "scientific" view of temporal matters, time is a kind of cosmic wastebasket, sitting there waiting for events to be deposited into it. But it's also clear, I think, that there is a corresponding view of time as the manifestation of relations between events. And it is this vision we want to explore here.

In the event-oriented picture of time, the events are a priori in our experience, just like the objects that have primacy of position in the non-Newtonian conception of space. By this view, we have to reject the idea that time is like some continually flowing stream of moments waiting to be filled by events. What's needed instead is the version espoused by Aristotle, in which "time is the measure of change with respect to before and after," where we understand *change* to be the experience of some structure of events. Let's see how all these Newtonian, Aristotelian and Leibnizian ideas fit into the simplicial complex framework we've been exploiting in this chapter.

The Newtonian view of time can be represented by the diagram in Figure 5.11. Here the numbered vertices represent specific measured moments of time, while the lines correspond to time intervals between the moments. This is a particularly primitive simplicial complex, one having an infinite set of vertices. We can attach a number to every vertex by associating each vertex with a particular measured moment as read off from, say, a stopwatch or a clock. This process of attaching numbers to the vertices creates what's technically termed a *pattern* on the 0-simplices. Since this pattern is associated with a set of zero-dimensional objects, let's give it the name τ_0. So this pattern is simply a rule by which we attach a definite number to each zero-dimensional object in the complex. Similarly, the numbers assigned to the one-dimensional edges joining the vertices form another pattern, τ_1, which represents the time interval between successive measured moments. Thus, in Newton's world the overall time pattern τ is what's called a *graded pattern*, which we represent symbolically as $\tau = \tau_0 \oplus \tau_1$.

The diagram in Figure 5.11 makes it clear why we think of Newtonian time as a *linear* concept associated with a complex: it consists of a set of "lines" (1-simplices) connected by "points" (0-simplices). When we use the Newtonian time axis to represent a set of observed real-world events, we try to produce somehow a "clock" whose time moments (the vertices) can be put into one-to-one correspondence with the set of events. The pattern τ_0 describes the NOW events, while the pattern τ_1 describes the interval pattern separating these events.

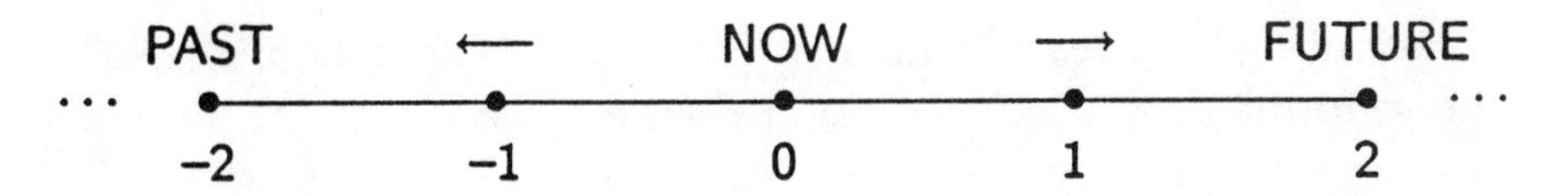

FIGURE 5.11. THE NEWTONIAN TIME PATTERN

In his book *In Search of the Miraculous*, P. D. Ouspensky presents a table of time intervals that he claims are the most crucial for humankind. These intervals are "the quickest impression," "breath," "waking and sleeping" and "life." According to Ouspensky, these intervals have the values 0.0001 seconds, 3 seconds, 24 hours and 80 years, respectively. All these intervals are, of course, measured by the scale of objective Newtonian time as outlined above. We'll see in a moment how they match up to the subjective time scale generated by taking a multidimensional view of time.

From the discussion of hierarchies in an earlier section, we know it's important to distinguish between events at level N, level $N+1$ and so on. Consequently, at any one of these levels we can expect events to generate a connective structure that we will term the *backcloth* for that level. It's with respect to this backcloth that we experience our sense of time. Let's see why.

Consider the backcloth at level N, which we denote $S(N)$. The structure of $S(N)$ will normally consist of 0-simplices representing those "events" that are points, 1-simplices characterizing the events that are pairs of points (edges) and, in general, p-simplices symbolizing those events formed by $p + 1$ points. If an event in this backcloth needs a p-simplex for its representation, let's agree to call it a *p-event*. This will correspond to a p-dimensional NOW event. In that case, the corresponding time interval needs to be a $(p + 1)$ event since such events are what join p-events together.

To illustrate what we mean by a p-event, consider the event "house." We only recognize the occurrence of this event when the entire house is complete and ready to be lived in. But many things have to take place in order for this to happen. For the sake of definiteness, let's say that for the house to be complete it needs to have a foundation, walls, floors, roof and exterior trim. So there are five vertices here, all of which taken together constitute the occurrence of the event "house," and the event "house" requires a 4-simplex for its representation, the vertices of which are the points "foundation," "walls" and so forth. Thus, recognition of the existence of the event "house" is a 4-event.

Note the important distinction between thinking of, say, a 1-simplex as a 1-event and regarding the same simplex as a 1-interval between two successive 0-events. In the latter case, we can order (in the linear Newtonian sense) the endpoints of the 1-simplex, while in the former situation we cannot. So when we think of a p-event, we are saying that the $p + 1$ points making up this p-simplex may not be able to be ordered in this way. All we can say is that they form a distinct, irreducible unit. The associated time intervals then become an experience of time passing, allowing us to move from one such p-event to the next p-event.

When we try to pretend that these time experiences are Newtonian, implying that every event is a 0-event, we are effectively treating the geometry of the backcloth $S(N)$ as if it were the linear structure of Figure 5.11. Intuitively, this corresponds to trying to warp the natural geometry of the events. So by forcing our time experience into the clock-time of Newton, we end up subjecting ourselves to structural stresses and strains that give rise to expressions like "time flies," "time drags" and "time is running out." Let's look a little deeper into this relation between experienced time and Newtonian clock-time.

For a Newtonian clock-time observer, every time interval has to be described in terms of intervals between events (i.e., in terms of edges, or 1-simplices). How would such an observer see a p-event in the Newtonian backcloth $S(N)$? The number of edges in a p-simplex is $p(p + 1)/2$. So in terms of the Newtonian clock-time unit, be it 1 second, 1 day or whatever, such a p-event would require *at least* this many units to manifest itself as an actual occurrence. So, for example, if the unit for our Newtonian observer is 1 day, a 7-event would appear to take $(7 \times 8)/2 = 28$ days to "arrive." We can term this number the *consolidation time* for a 7-event. But it may require more units since the event can only be seen to have happened when the observer has gone around the edges of the 7-simplex in some order. This Newtonian observer might choose a path through these edges that involves traversing some edges several times before all the edges of the 7-simplex have been covered.

Notice now that if we allow our observer to move up the hierarchy from level N, say, to level $N + 1$, the observer would be able to see all the faces of the p-simplex (i.e., all the subevents of the p-event). This observer would need a Newtonian clock-time for each of them. In the case of a 2-event, for instance, the $N + 1$-level observer would see six subevents: three 0-events (the vertices of the triangle) and three 1-events (the edges of the triangle). In general, it can be shown that this *super consolidation time* for a p-event is given by the formula $p \times (p + 1) \times 2^{p-2}$. Thus, for a 7-event, the super consolidation time is 1792 clock-time units. So the level-N observer sees a 7-event occur in 28 days; for the level-$N + 1$ observer, the same event takes 1792 days, or approximately 5 years. This kind of analysis focuses on the events. Now let's take a moment to consider the time intervals between events.

We have already noted that the intervals separating p-events require $p + 1$ points. Consider first a Newtonian observer at level N. Assume such an observer sees only the edges connecting two p-events. Figure 5.12 shows this situation for a pair of 2-events. Then the final interval between the "end" of one p-event and the end of the next is formed by counting each of these edges once, then adding that

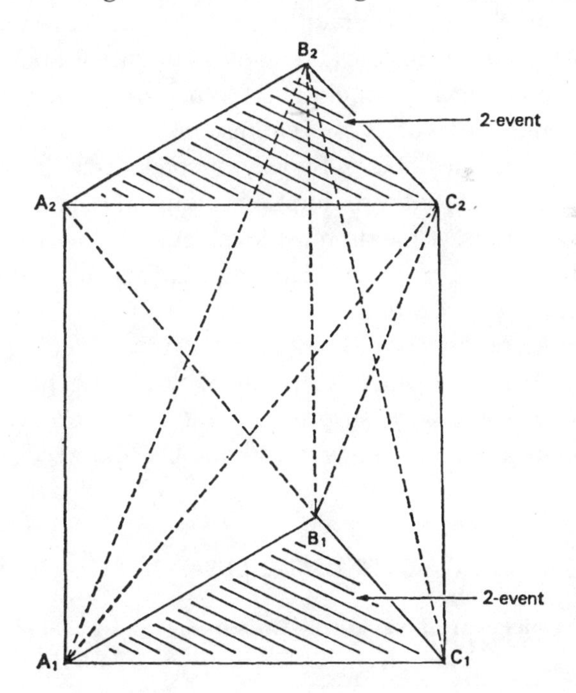

FIGURE 5.12. SUCCESSIVE 2-EVENTS WITH CONNECTING EDGES

number to the consolidation time. This gives a total time of $(p + 1)^2 + p(p + 1)/2$ time units. Again focusing on the case $p = 7$, we see that the N-level time interval between two successive 7-events will be 92 units of Newtonian clock-time.

Taking an $N + 1$-level view, we need to move from the first p-event to each *subset* of the vertices of the second p-event. Thus, we add the total of these edges to the previous consolidation time, giving us the formula $(p + 1)^2 \times 2^p + p(p + 1) \times 2^{p-2}$ for the time interval between the p events as seen by an $N + 1$-level Newtonian observer. Referring to Figure 5.12, this would mean that such an observer would have to count the edges A_1A_2, A_1B_2, A_1C_2 and then the pairs of edges $\{A_1A_2, A_1B_2\}$, $\{A_1A_2, A_1C_2\}$, $\{A_1B_2, A_1C_2\}$ and then triples of edges $\{A_1A_2, A_1B_2, A_1C_2\}$ and then repeat this process starting with B_1 and C_1 in turn. The final consolidation time would then be formed by adding to this count the super consolidation edge count for the second 2-event. In this higher-level view, the interval between successive 7-events turns out to be 9984 time units. This means that if an $N + 1$-level Newtonian observer saw an interval of 1 day between two 0-events, it would take nearly 30 years for this observer to see a pair of 7-events pass.

These speculative sums contain many assumptions about the relation between hierarchically experienced time and clock time. Oddly enough, though, they are very reminiscent of Ouspensky's "times of the cosmoses" reported above. By way of comparison, Table 5.1 shows Ouspensky's values for these durations alongside the time intervals of successive 8-events at hierarchical levels $N - 2$, $N - 1$, N and $N + 1$, taking 24 hours as a given standard in both lists and calling this level N. The similarities are striking.

By now, the reader should have no difficulty in seeing the intimate connections between our perception of time and the notion of an event being in some way "surprising." But in the vernacular, surprise is usually associated in some way with the likelihood of an event (i.e.,

TABLE 5.1.
OUSPENSKY'S TIME INTERVALS VERSUS TIME DURATION BETWEEN 8-EVENTS

Ouspensky	**Quickest Impression** 10^{-4} seconds	**Breath** 3 seconds	**Waking/Sleeping** 24 hours	**Life** 80 years
q-analysis	**Level N – 2** 1.34×10^{-4} seconds	**Level N – 1** 3.4 seconds	**Level N** 24 hours	**Level N + 1** 70 years

the probability of its occurrence). So in the next section we develop this theme a bit further, showing not only how surprise can be rigorously formulated in terms of connective structure in a complex of events, but also how classical probability theory arises from the consideration of a very special kind of maximally connected complex.

SOME SURPRISING CONNECTIONS

As Albert Einstein and Erwin Schrödinger were strolling along Berlin's Unter den Linden one day in 1926, Einstein remarked, "Of course, every new theory is true, provided you suitably associate its symbols with observed quantities." Here Einstein was referring to the epistemological difficulties surrounding Schrödinger's then-recent development of what has come to be called the wave function interpretation of quantum phenomena.

In this conventional view of quantum phenomena, every observable aspect of a material object—position, momentum, spin, charge, energy—is encoded into a single mathematical gadget, the wave function. So the quantum mechanician's prescription for predicting the results of an experiment designed, say, to measure the momentum of a speeding electron is quite simple. First construct the wave function for this experimental situation, which involves solving what's called the *Schrödinger equation*. Then "slice" this solution in a well-defined mathematical way, the particular slice corresponding to the attribute you want to measure—in this case the electron's momentum.

While Schrödinger's equation for the quantum wave function serves admirably as a vehicle by which to predict the outcome of laboratory experiments, the wave function itself has defied all attempts to give it an interpretation in terms of physically observable entities. It remains today as much of an ontological mystery as it did the day Schrödinger first wrote it down. In fact, Schrödinger himself became so exasperated with Niels Bohr's persistent attempts to get him to admit that his wave function had no physical interpretation that he once blurted out, "I am sorry I ever started to work on atomic theory." Strong words from the man who in 1933 was awarded the Nobel Prize in Physics for "new insights into atomic theory."

Probably the biggest mystery of the quantum world unveiled by Schrödinger, Bohr, Heisenberg and others in the 1920s is what we now call the *quantum measurement problem*. As noted, the values of attributes like position, momentum and spin of quantum objects like photons and electrons are all described by the Schrödinger wave function. But until a measurement of one of these attributes is actually made, the wave function describes merely the likelihoods of the possible outcomes of such a measurement. Once the measurement is actually taken, of course, the potential outcomes are replaced by a single outcome, which we term the *result* of the measurement.

The problem here is that prior to the measurement, the wave function exists as a kind of mathematical wave of probability. Yet as soon as we make an observation, this wave "collapses" to a single point—the outcome of the measurement. The essence of the measurement problem is to ask how this collapse comes about and what it actually means in physical terms. In short, what's so special about the act of a measurement? And how does this *physical* action of observing a system collapse what appears to be a purely *mathematical* gadget, the wave function? Although the point is never mentioned in courses on probability theory, it's of considerable interest to note that the very same kind of collapse also occurs in classical probability theory.

Probabilities and Simplices

One of the great challenges to both science and philosophy is to provide a rational, coherent account of the perceived uncertainty surrounding the events of daily life. Classical probability theory offers one such approach but is riddled with many well-known epistemological flaws and paradoxes. The theories of fuzzy sets, satisficing and possibilities represent recent attempts to rectify some of the deficiencies in the classical methods. Each of these newer schemes has at its heart the basic fact that randomness is only one face of the mask of uncertainty. Actually, most of the uncertainty we experience about everyday events cannot usually be attributed to the influence of random mechanisms, at all. Rather, it seems to stem from an inherent vagueness, or lack of information, either in the linguistic description or in other circumstances surrounding the situations we find ourselves confronting. Here we want only to indicate the manner in which the multidimensional structure of a simplicial complex allows us to formalize many of the notions of uncertainty, probability and surprise in a manner providing some insight into the basic difficulties involved in characterizing uncertainty. We'll then give a few suggestions for how a coherent theory of uncertainty and surprise might be developed.

Consider an experiment in which we toss a fair die four times in succession. Suppose our interest is in whether or not the face 6 appears. Let's label the elementary events (i.e., the outcome) of each throw by x_1, x_2, x_3 and x_4. Thus, the quantities x_i are what probabilists call *random variables*. We can then combine these individual elementary events into a *set* of elementary events denoted by X. To be more specific, the element x_1 represents the event that 6 occurs on the first toss, x_2 is the event that 6 turns up on the second toss and so on. Now consider the new set Y, consisting of *compound* events associated with the entire experiment of four tosses. In this experiment there are sixteen possible outcomes ranging from no occurrences of a 6 to all tosses resulting in 6. So we label the elements of Y as

$$Y = \{y_0, y_1, y_2, y_3, y_4, y_{12}, y_{13}, y_{14}, y_{23}, y_{24}, y_{34}, y_{123}, \\ y_{124}, y_{134}, y_{234}, y_{1234}\}$$

where y_0 means no 6 occurred, y_2 means that a 6 occurred only on the second toss and so forth.

If we take the elements of X to be the vertices of a complex, letting the elements of Y be simplices formed from these vertices, it's easy to develop a relation R_Y linking these two sets. It is defined by the rule

> "The pair of elements (y, x) is in the relation R_Y if and only if elementary event x forms part of the compound event y."

We can easily compute the incidence matrix for R_Y: it has a 1 in the column labeled x_i if and only if the integer i appears as one of the index numbers on simplex y. For example, (y_{13}, x_1) is in R_Y but (y_{13}, x_2) is not.

Computing the chains of q-connection in this complex, we find that the complex has only a single component at each dimensional level. In fact, it's easy to see that what we're dealing with here is the *single* simplex y_{1234} and all of its faces. This is exactly the kind of structure for which classical probability theory works well for expressing our sense of the unknown and the uncertain. Let's see why.

The complex whose simplices are the compound events y_i represents what the probabilist terms the *sample space* of the die-tossing experiment. But in contrast to the usual view of events as dimensionless objects, the multidimensional view distinguishes strongly between the compound 0-events (y_1, y_2, y_3 and y_4), the compound 1-events (y_{12}, y_{13}) and so on up to the single 3-event (y_{1234}). *Before* the experiment is performed, our sense of the likelihood of the outcome is measured by attaching numbers (probabilities?) to each simplex. *After* the experiment has been carried out, these numbers have rearranged themselves throughout the complex; all simplices now have the value 0 except for that single simplex corresponding to the actual outcome. By convention, we set the value of that simplex equal to 1.

So we see that carrying out the experiment corresponds to some kind of "traffic" on the complex. But traffic of this sort can move freely about from one simplex to another only if the complex is sufficiently richly connected at all dimensional levels to support such a free flow of dimensionally significant numbers. Basically, this means that the complex must be free of dimensional obstacles at all dimensions. In other words, it must have a single chain of connection for all dimensional levels—just as with the die-tossing complex above. Cases from classical probability theory, when the complex in question consists of a single simplex and all of its faces, are the simplest examples of situations when this will occur.

In connection with the die-tossing experiment, the probabilist would attach the following a priori values to the elements of Y:

$$E(y_0) = \frac{625}{1296} \qquad E(\text{0-simplices}) = \frac{500}{1296} \qquad E(\text{1-simplices}) = \frac{150}{1296}$$

$$E\,(\text{2-simplices}) = \frac{20}{1296} \qquad E\,(y_{1234}) = \frac{1}{1296}$$

These numbers express the probabilist's sense of the likelihood of events and are formed by recognizing that there are 1296 possible outcomes of the four rolls of the die. Of these, there are 625 sequences containing no 6s, 500 sequences showing a single 6 and so down to a single sequence in which all four rolls are 6s. Invoking the fact that the rolls are independent leads to the probabilities stated above. After the experiment is over, these numbers have redistributed themselves so as to coalesce on that one simplex corresponding to the actual outcome. But this rearrangement is possible only if the numbers associated with p-events can freely move about and "reaffiliate" themselves with events at all levels of connectivity p. This can happen only in structures having a single component at each connectivity level.

The main point about connectivity here is that if the numbers assigned to the compound events in Y are to represent our sense of the likelihood of the outcome of the experiment, then they must do so both before *and* after the experiment. But this requires the kind of free flow of traffic discussed above, a flow that is possible only if the complex representing the events is fully connected at all levels. Thus we conclude that classical probability theory will in general reflect our sense of the likelihood of events only for those structures possessing a single component at each dimensional level.

So we see that the redistribution of numbers over the complex when we actually perform the experiment is completely analogous to the collapse of the Schrödinger wave function in quantum mechanics. In most versions of quantum theory, an object's attributes exist as *potentials*, the various possibilities weighted according to the probability distribution specified by the solution to the Schrödinger equation. This probability distribution is what we call the wave function. Following the actual measurement, a definite value is obtained for any given attribute, the wave function then being said to have collapsed to the single value actually observed. All other possibilities then have probability zero.

As a result, our complex of events, together with its associated likelihood numbers, is analogous to the quantum-mechanical wave function. But it has the additional feature that the possible events (simplices) have a dimensional character that must be respected when considering any redistribution of likelihoods following an experiment (observation). The quantum-mechanical implications of these dimensional factors have not been investigated as yet, classical quantum theory having confined itself to the same case as classical probability theory, viz., complexes with a single component at each dimensional level of connection. Now let's talk about surprise.

One of the principal uses of probability theory is to provide a numerical measure of our sense of how surprising the occurrence of a particular event would be. By the foregoing arguments, we find that the concept of surprise is intimately tied up with the connective structure linking events in the space of possible outcomes. In particular, to develop a decent theory of surprise we need a measure of the "reachability" of a q-event, call it σ_q, from another base event σ_p^* (the NOW) in the complex.

On intuitive grounds, the surprise value of a q-event σ_q should be a number that:

1. Reflects the level of connectivity between σ_q and the NOW event σ_p^*. In particular, if there is no chain of q-connection between the two, then the surprise value should be zero, since we can't be *surprised* if the event σ_q cannot be experienced from σ_p^* if there is no appropriate dimensional path to move from the NOW event to σ_q.
2. Is greater if there are a large number of disjoint p-chains from the base event to σ_q, since it is more surprising if a large number of p-chains go between the two events than if there are only a small number of such paths.
3. Is smaller if the dimension of the NOW event is large since it is less surprising that q-chains exist from the base event to σ_q if the NOW event has more q-dimensional faces.

One measure satisfying these conditions involves the number of disjoint q-dimensional chains linking the base event to σ_q. For the more mathematically inclined, the specifics of this measure are given in the To Dig Deeper section.

An interesting and timely example of the use of surprise theory arises with the surprise value of a technological disaster, like that associated with nuclear power plants such as Three-Mile Island or Chernobyl. Let the vertices X of the complex represent various technological features of the plant. These elements might be things like the position of control rods, the level of coolants and the pressure in regulators. Let the set of simplices Y forming the complex K that represents the plant consist of combinations of features that we term properties or behaviors of the plant.

If all the vertices are initially in the state OK, then we say that all is well. Assume that during the course of operation of the plant some vertices turn into antivertices—their OK activity turns into not-OK. So the complex K turns into a new complex K^1. As the process of vertices shifting to and from OK $\leftrightarrow$ not-OK unfolds, we have the progression

$$K \rightarrow K^1 \rightarrow K^2 \rightarrow \ldots \rightarrow K^D$$

where the event σ = DISASTER belongs to K^D. We can now ask the question, Given the base event (state) σ_p^* in K, what is the surprise value for an event σ in K^D? Clearly, we would like to arrange our technology to make this number large.

On the perhaps optimistic note that such a technological dream can be transformed into a reality, let's leave the idea of surprise arising as the result of multidimensional connections, turning our attention to a different way in which connective structure can lead to the unexpected: self-similarity in both space and time. This comes about via the twin phenomena of self-organization and emergence.

SIX

THE EMERGENT

Intuition: Surprising behavior results only from complicated, hard-to-understand interactions among a system's component parts.

The notion of structure is comprised of three key ideas: the idea of wholeness, the idea of transformation, and the idea of self-regulation.

—JEAN PIAGET

God has put a secret art into the forces of Nature so as to enable it to fashion itself out of chaos into a perfect world system.

—IMMANUEL KANT

Large streams from little fountains flow,
Tall oaks from little acorns grow.

—DAVID EVERETT

CHECKERBOARD COMPUTERS

In the summer of 1967 I found myself working as a programmer at the RAND Corporation in Santa Monica, California. One day my boss dropped by my office to say that there was a Harvard professor visiting for the summer who needed a little programming help, and could I deal with the situation. While my enthusiasm for helping Harvard professors wasn't much greater then than it is now, in those days RAND was a place where even a summer consultant from Harvard might be doing something interesting. So I agreed to get in touch with this guy in the morning and see what his problem was. As fate would have it, this "guy" turned out to be the world-renowned political economist Thomas C. Schelling, who was engaged in an experiment that led directly to my first professional encounter with the phenomenon of *emergence.*

The next day Schelling told me that to while away a January blizzard the previous winter, he had pulled a checkerboard off the shelf at his vacation home in the White Mountains of New England to test a wild theory he had concocted about the dynamics of racial patterns in housing. Specifically, what he was out to explore was the matter of whether decisions made by individuals about where to live, decisions based upon simple rules employing only local information (i.e., information only on the color of one's immediate neighbors), could generate observable patterns of racially segregated housing in initially mixed neighborhoods. This was a phenomenon Schelling termed *tipping*, which to my physics-trained eye looked like a sociological version of the kind of phase transition that takes place when water turns to ice. He said that he had experimented with the idea by placing black and white pieces on a checkerboard, finding that some rules for shifting the pieces from one "homesite" to another did lead to such racial tipping. But he wanted to do a more systematic set of experiments using much larger checkerboards (urban areas) and test a broader set of rules. Enter the programmer and his computer.

To respect the rather puny memory available on the computers of the day, Schelling and I partitioned the urban region into a rectangular grid of 16 by 13 cells, assuming that each cell represented a homesite that could be occupied by a white or a black family or be empty. Thus, there was a total of $3^{(16\times13)} = 3^{208} \approx 10^{99}$ possible states this urban area could be in, each of which represented one housing pattern distribution of the black and white families in the region.

As the rule of moving, Schelling postulated that both the blacks and the whites would prefer to have a certain percentage of their immediate neighbors be of the same group. So if this were not the case for a particular distribution, the model assumed that each person would move to the nearest available location where the percentage of like neighbors was acceptable. In order to have a reasonable choice of where to move, the earlier hand-done checkerboard experiments showed that around 25 to 30 percent of the housing locations should be left vacant.

Defining the neighbors of an individual homesite as all 8 immediately adjacent locations, and starting with the racially mixed initial state shown in Figure 6.1, in which the symbol o denotes a white family, the symbol # denotes a black household and a blank space means the housing space is empty, we came to the steady-state distribution of Figure 6.2(a) using the condition that at least half of one's neighbors must be of the same color. Figure 6.2(b) shows the resulting steady-state distribution from the same initial configuration when the requirement is that at least one-third of one's neighbors must be of the same group. Notice the "tipping" that takes place as the area changes from racially mixed to racially segregated. Although neither Schelling nor I realized it at the time, this racial integration experiment is a perfect example of a mathematical gadget that Stanislaw Ulam had already invented under the rubric of a *cellular automaton* (CA).

Cellular Automata

Consider an infinite checkerboard in which each square can be colored either black or white at each instant of time *t*. Assume there is a rule specifying what the color of each square should be as a

```
o # # # # o o o o     # #   o o
o   # o o o   # #     #   # o
#   # o o # #   #     o #   # #
#     # #     o # # # o o
o   o o # # # #     #   #   o o
  # o #   o o #     o o     # #
# o o #         o o o # # #
o   # o   # #   # o o o       #
o   # o         # # o         #
o o       #     o # o o o o # #
  o # # o o o o   o # #   o # #
#   o # o #   o o # o # o   o
  o o     o # o # o o o     # #
```

FIGURE 6.1.
INITIAL HOUSING DISTRIBUTION

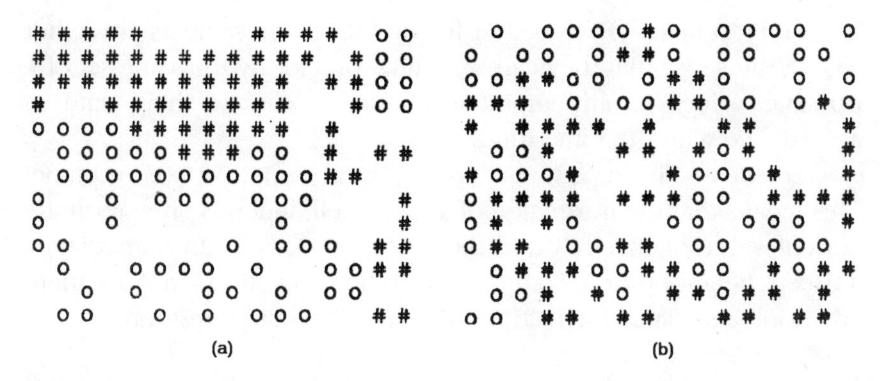

FIGURE 6.2. (A) HALF OF NEIGHBORS THE SAME COLOR; (B) ONE-THIRD OF NEIGHBORS THE SAME COLOR

function of the colors of its immediate neighbors. Now let an initial pattern of black and white squares be given at time $t = 0$. Turning the system on and letting the rule of state transition operate, we examine the configurations of black and white squares that emerge as time unfolds in discrete steps $t = 1,2,\ldots$. In other words, we watch the squares blink on and off at each time step in accordance with the dictates of the rule of state transition, looking for whatever patterns might turn up. This setup describes the prototype for a cellular automaton (CA) or, more accurately in this case, a two-dimensional CA, since the playing field (i.e., the state space) here is a planar grid. Let's consider for a moment the key ingredients constituting such a computational and mathematical object.

- *Cellular state space*—The backcloth upon which the dynamics of the automaton unfold is a cellular grid of some kind, usually a rectangular partitioning of ordinary one- or two-dimensional space. Our interest here will focus upon these classes of cellular automata for the simple reason that they are easy to visualize and capture most of the features of higher-dimensional CA.
- *Finite states*—Each cell of the state space can assume only a finite number k of different values. So if we have a finite grid of N cells, the total number of possible states is also finite, equaling k^N.
- *Deterministic*—The rule fixing the value of each cell at a given moment is a deterministic function of the current value of that cell and the values of the cells in a neighborhood of that cell. On occasion people modify this condition by allowing stochastic transitions. But for our purposes here, we consider only deterministic transition rules.

• *Homogeneity*—Each cell of the system is the same as any other cell, in the sense that they can each take on exactly the same set of k possible values at any moment and each change their state in accordance with the same rule.

• *Locality*—The state transitions are local in both space and time. This means that the next value of a given cell depends only upon the current value of that cell and the values of cells in an immediately adjacent neighborhood. So there are no time-lag effects, nor are there any nonlocal spatial interactions affecting the state transition.

For one-dimensional CA, the local neighborhood of a given cell consists of a finite number of cells on either side of that cell; for two-dimensional CA, there are traditionally two basic neighborhoods of interest: the *von Neumann neighborhood*, which consists of those cells vertically and horizontally adjacent to the given cell, and the *Moore neighborhood*, which also includes those cells that are diagonally adjacent. These neighborhoods are illustrated in Figure 6.3. As it turns out, most of the principal features of cellular automata that make them both theoretically and practically interesting can be seen by focusing attention on the simplest class of such objects, the one-dimensional cellular automata (1-D CA).

A 1-D CA consists of an infinite string of cells changing values according to a given rule. We can think of some cosmic clock ticking away such that at each tick every cell in the string assumes a value determined by its previous value and the previous values of cells in its neighborhood. Therefore, a 1-D CA is specified by two numbers, let's call them k and R, together with a rule determining the next value of each cell. The first number, k, specifies how many values are possible for each cell, while the quantity R refers to the size of the

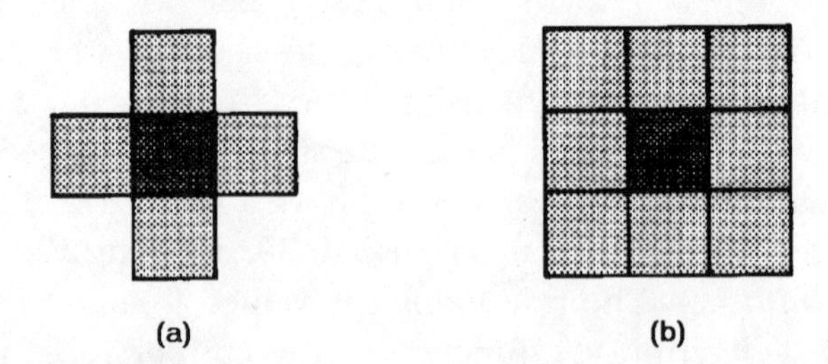

FIGURE 6.3. (A) VON NEUMANN NEIGHBORHOOD; (B) MOORE NEIGHBORHOOD

neighborhood used to compute the next value of a cell. So, for example, when $k = 2$ and $R = 1$, a particular rule might specify that if a string of three successive cells has the values 011, the next value of the center cell would be 1.

The set of rules that define a given 1-D CA must uniquely determine the fate of a cell for every possible set of cell values in its neighborhood. Clearly, if k and R are large, the number of possible rules is enormous. Even in the simplest case, when $k = 2$ and $R = 1$, there are 256 possible rules, each of which corresponds to a different 1-D CA. Just to get a feel for these objects, let's look at the behavior of a 1-D CA defined by one of these 256 rules, something called the *Mod 2 Rule.*

The local rule for the time evolution of this Mod 2 CA can be represented by the diagram

$$\frac{111}{0}\ \frac{110}{1}\ \frac{101}{0}\ \frac{100}{1}\ \frac{011}{1}\ \frac{010}{0}\ \frac{001}{1}\ \frac{000}{0}$$

Here the upper part shows the eight possible states that a row of three successive cells can be in at time t, whereas the lower half shows the value that the central cell of each trio assumes at time $t + 1$. This is called the Mod 2 Rule since the value of each cell is determined by simply adding the value of its two neighbors, dividing that number by 2 and keeping the remainder.

The following diagram shows the action of the Mod 2 Rule over one time step:

$$\frac{(\text{time } t)\ \ldots 1\,0\,1\,1\,0\,1\,1\,0\,1\,0\,1\,0\,1\,1\,0 \ldots}{(\text{time } t+1)\ *\ *\ 0\,0\,1\,1\,0\,1\,1\,0\,0\,0\,0\,0\,1\ *\ *}$$

To see how complicated the behavior of some 1-D CA can be, Figure 6.4 shows the evolution over several time steps of four different rules. In each case, the initial state of each cell is chosen to be 0 or 1 with equal probability. To make it easier to see the patterns, cells in state 0 are left blank in the figure, while cells in state 1 are colored black. Note also that time is taken to move downward, so that the top line in the figure is the configuration at time $t = 0$, the second line is the CA state at $t = 1$ and so forth.

The patterns of Figure 6.4 illustrate the four possible types of long-run behavior, which Stephen Wolfram has termed Types A, B, C and D. Recalling the discussion of attractors given in the opening

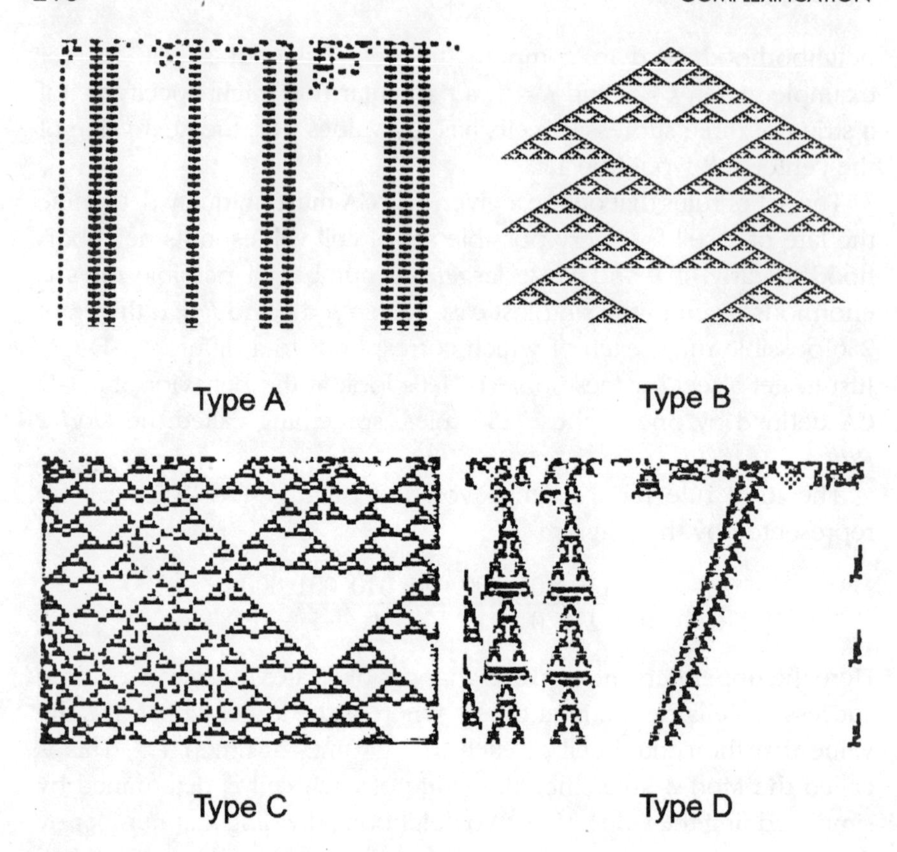

FIGURE 6.4. THE EVOLUTION OF FOUR DIFFERENT 1-D CA RULES

chapter, the reader will have no difficulty recognizing these CA behaviors as being the counterparts in discrete time and space of the attractor types we saw in earlier chapters for continuous-time, continuous-state dynamical systems. Here are the correspondences:

Attractor Types	
Dynamical System	Cellular Automata
Fixed point	Type A
Limit cycle	Type B
Strange attractor	Type C
Quasiperiodic orbit	Type D

These kinds of simple CA have been used to model a bewildering variety of processes ranging from the sequences of nucleotide bases on a strand of DNA to the complexity of both human and computer languages. To see how such applications go, let's briefly indicate how 1-D CA can be employed to capture some aspects of the development of a plant.

The Growth of Plants

In the late 1960s, Dutch biologist Aristid Lindenmayer proposed a CA model for the development of filamentous plants, such as the blue-green algae *Anabaena*. His model contains the novel feature that the number of cells is allowed to increase with time according to a recipe laid down by the state-transition rule. In this way the model "grows" in a manner mimicking the growth of a filamentous plant.

The simplest version of one of Lindenmayer's *L-systems* involves a one-dimensional grid that starts at time $t = 0$ with a single active cell having value 1. The state-transition rule is as follows: "The next value at cell i is given by its current value together with the value of the cell immediately to its left (i.e., cell $i - 1$)." Thus we can express the complete state transition at cell i as

$$(i = 0,\ i - 1 = 0) \Rightarrow i = 0 \qquad (i = 0,\ i - 1 = 1) \Rightarrow i = 1$$
$$(i = 1,\ i - 1 = 0) \Rightarrow i = 11 \qquad (i = 1,\ i - 1 = 1) \Rightarrow i = 0$$

The interesting feature of this rule is the possibility for cell division that occurs when $i = 1$, $i - 1 = 0$. If we take an initial state such that the first cell at the left has value 0, while the cell next to it has value 1, then the first four state transitions are 01, 011, 0110, 011011. Already we see here the growth of the initial "seed" at cell 2 to several seeds at cells 2, 3, 5 and 6.

It must be admitted that these linear strings of 0s and 1s don't look much like what anyone would even charitably call a plant. So to bring out more clearly the connection between L-systems and the real world of plants, let's soup up our notation by extending the symbol alphabet to include the new symbols [and]. While we're at it, let's also jazz up the state-transition rule so that the four symbols of our system always transform in the following way:

0 → 1[0]1[0]0
1 → 11
[→ [
] →]

To see how this rule works, suppose we start with the string consisting of the single symbol 0. With the above transformation rules, the first three steps of this CA turn out to be

$t = 0$: 0
$t = 1$: 1[0]1[0]0
$t = 2$: 11[1[0]1[0]0]11[1[0]1[0]0]1[0]1[0]0

This still doesn't look much like a plant. But we can convert strings of this type into a treelike structure by treating the symbols 0 and 1 as line segments while regarding the left and right brackets as branch points.

One way to get something that looks faintly plantlike out of all this is to leave the 1-segments bare while placing a leaf at the end of each 0-segment. So, for instance, if we have the string 1[0]1[0]0, its stem consists of the three symbols not in brackets. These are a 1-segment beneath another 1-segment, which in turn is topped off by a 0-segment. Two branches, each with a single 0-segment, sprout from this string. The first branch is attached above the first segment, while the second branch occurs after the second segment. As for the direction of the branch, the simplest convention is to specify that for any given stem, the branches shoot off alternately to the left and to the right. Figure 6.5 shows the first three generations of a plant obtained by using the L-system grammar given above, together with the foregoing rule regarding leaves and branches. The reader will find much more realistic examples of how L-systems can mirror the development of plants in the references cited in the To Dig Deeper section.

With the idea of rules of this sort allowing a CA to change its configuration as it goes along, it's only a small step to the consideration of whether lifelike objects might be created on the grid of appropri-

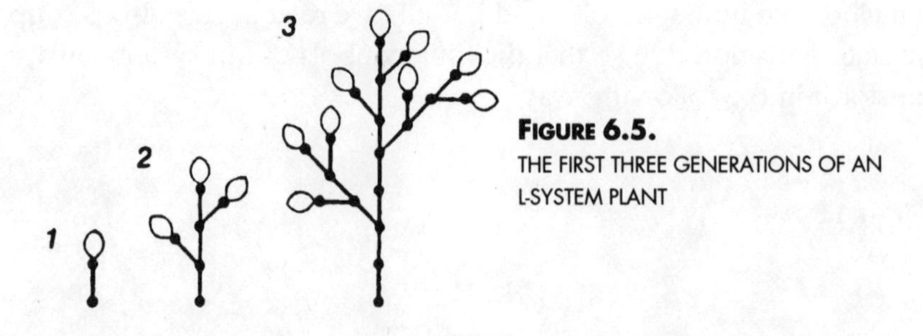

FIGURE 6.5.
THE FIRST THREE GENERATIONS OF AN L-SYSTEM PLANT

ately specified CA. Oddly enough, this was the question that intrigued John von Neumann nearly fifty years ago. And he developed many of the properties of CA to answer it.

THAT'S LIFE?

In a lecture at Caltech in 1948, John von Neumann used the idea of a CA to lay down the conditions that any entity would have to satisfy if it were to be capable of building a copy of itself (i.e., self-reproduction). What concerned von Neumann was the kind of logical organization and functional activities an object would have to possess to be able to build a copy of itself. Thus, von Neumann wanted to abstract from the processes of self-reproduction in nature the *logical* form of the reproduction process—independent of its realization in any particular material structure.

To solve this problem, von Neumann created a universal Turing machine consisting of a 2-D CA having 29 states per cell. His idea was to represent the initial machine as a particular pattern in this CA array. Self-reproduction would then be said to have occurred if a rule of state transition (using the 5-cell von Neumann neighborhood) could be found that would cause the initial pattern to be duplicated elsewhere in the array. But for genuine self-reproduction, the instructions for building the copy have to be contained in the initial configuration itself. We'll return to this point in a moment.

Von Neumann showed how his 29-state CA could be capable of universal *construction*, from which self-reproduction followed as a special case when the machine described on the constructor's input was the constructor itself. Unfortunately, the details of von Neumann's argument are far too technical to enter into here. But the curious reader can find the complete story in the items cited in the To Dig Deeper section.

The key to the self-reproduction problem turns out to be the way we handle the issue of copying the "blueprint" of the machine. Suppose we've succeeded in building a universal constructor. We then feed the plans for the constructor back into it as input. The constructor will then reproduce itself. But it will not reproduce the instructions describing how to build itself. This is a trivial and nonperpetuating type of reproduction, and not at all what we have in mind when we speak of a self-reproducing machine. How do we arrange it so that the blueprint, as well as the constructor, is faithfully reproduced? This was the big difficulty that von Neumann had to surmount. His solution involved using the information on the blueprint in two completely different ways.

Von Neumann's way out of the blueprint dilemma was to build a "supervisory unit" into the constructor. This unit functioned in the following manner: Initially the blueprint is fed into the constructor as before, and the constructor reproduces itself. At this point the supervisory unit switches its state from construction-mode to copy-mode and proceeds to copy the blueprint as raw, uninterpreted data. The copy is then appended to the previously produced constructor (which includes a supervisory unit), and the self-reproducing cycle is complete. The key element in this scheme is to prevent the description of the constructor from becoming a part of the constructor itself (i.e., the blueprint is located outside the machine and is then appended to the machine at the end of the construction phase by the copying operation of the supervisory unit).

The crucial point to note about von Neumann's solution is the way information on the input blueprint is used in two fundamentally different ways. It's first treated as a set of instructions to be *interpreted*. These instructions, when executed, cause the construction of a machine somewhere else in the CA array. Thereafter, the information is treated as *uninterpreted* data, which must be copied and attached to the new machine. These two different uses of information are also found in biological self-reproduction: the interpreted instructions correspond to the process of genetic *translation*, while the blind copying of the uninterpreted data corresponds to the process of genetic *replication*. These are exactly the processes involved in the operation of every living cell, and it's worth noting that von Neumann came to discover the need for these two different uses of information several years before their discovery by biologists working on the mysteries of DNA. The only difference between the way von Neumann set things up and the way nature does it is that von Neumann arbitrarily chose to have the copying process carried out after the construction phase, whereas nature copies the DNA early on in the cellular reproduction process.

As noted above, genuine self-reproduction has to take place in a very particular way. But there are many types of "pseudo" self-reproduction. A simple example is the CA defined by the Mod 2 Rule using the von Neumann 5-cell neighborhood in the plane. In this case, starting with a single ON cell, a little later we will see 5 isolated cells, each of which is also ON. Clearly, this doesn't fit with the everyday notion of self-reproduction, since the initial configuration was reproduced by the state-transition rule rather than by following a set of rules for reproduction contained within itself. So reproduction here is due solely to the transition rule built in to the "physics" of the environment and in no way resides within the configuration itself.

It turns out that simpler machines than von Neumann's can also be shown to be capable of self-reproduction. So the question arises, How simple can a self-reproducing machine be? This is the flip side of von Neumann's original question, which involved consideration of what would be *sufficient* for self-reproduction. Now we are concerned with what's *necessary*. As a concrete example of one such simple self-reproducing CA, let's look at the what is probably the most well-chronicled CA of all time.

The Game of Life

In the October 1970 issue of *Scientific American*, columnist Martin Gardner introduced the world to a simple board game that has come to be called the *Game of Life*. In actuality, *Life* is not really a game, since there are no players; nor are any decisions to be made. Rather it is a cellular automaton. In fact, in the terms we've been using in this chapter, *Life* is a 2-state, 2-D CA.

The rule of state transition for the *Life* CA was laid down by the game's inventor, Cambridge University mathematician John Horton Conway. It uses the Moore neighborhood (i.e., all 8 adjacent cells), specifying the following fates for the central cell of each neighborhood:

1. The cell will be ON in the next generation if exactly 3 of its neighboring cells are currently ON.
2. The cell will retain its current state if exactly 2 of its neighbors are ON.
3. The cell will be OFF otherwise.

This rule is Conway's attempt to balance out a cell's dying of "isolation" if it has too few living neighbors and dying of "overcrowding" if it has too many.

Figure 6.6 shows the fate of some initial triplet patterns using this

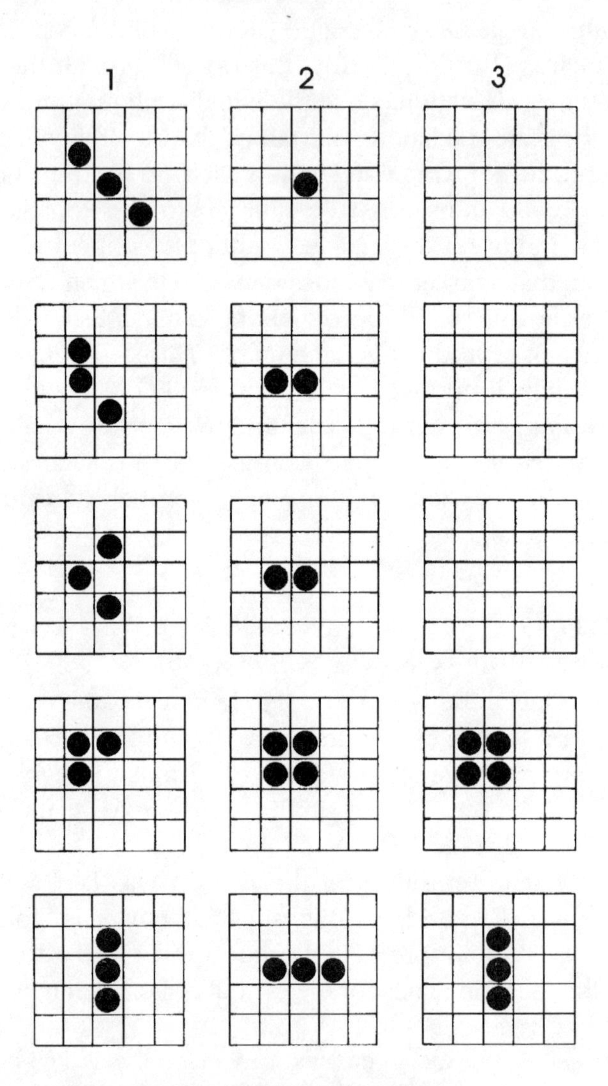

FIGURE 6.6. SOME TRIPLET HISTORIES IN THE *GAME OF LIFE*

rule. We see that the first three triplets die out after the second generation, whereas the fourth triplet forms a stable Block, and the fifth, termed a Blinker, oscillates indefinitely. Figure 6.7 shows the life history of the first three generations of an initially more-or-less randomly populated life universe.

Of considerable interest in the *Life* CA is whether there are initial configurations that are eventually copied elsewhere in the array. The

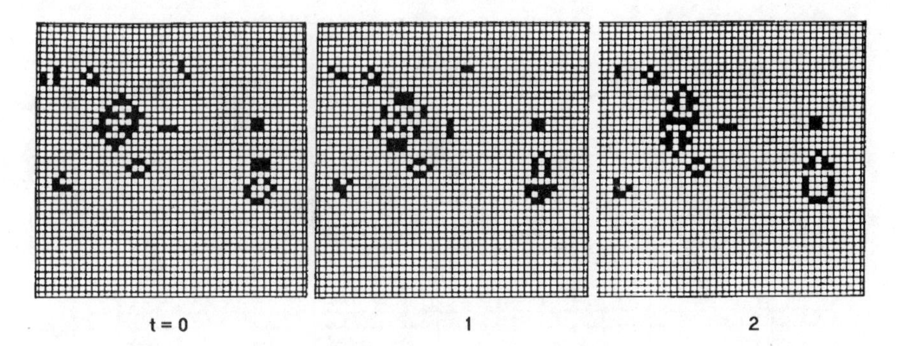

FIGURE 6.7. A TYPICAL *GAME OF LIFE* HISTORY

first example of such a configuration is the Glider, which is displayed in Figure 6.8.

In the early days of *Life*, researchers conjectured that because of the overpopulation constraint built into the CA rule, there were no configurations that could grow indefinitely. Conway offered a $50 reward to anyone who could produce such a configuration. The configuration that won the prize, termed a Glider Gun, is depicted in Figure 6.9. Here the Gun, shown at the lower left of the figure, is a spatially fixed oscillator that repeats its original shape after thirty generations. Within this period, the Gun emits a Glider, which wanders across the grid and encounters an Eater, which is shown at the top right of the figure. The Eater, a fifteen-generation oscillator, swallows up the Glider without undergoing any irreversible change itself. Since the Gun oscillates indefinitely, it can produce an infinite number of Gliders, implying that configurations that can grow indefinitely do exist.

While the *Life* rule has led to a veritable cornucopia of strange and captivating patterns, our concern here is with the issue of self-reproduction. So I'll leave it to the reader to consult the references cited in the To Dig Deeper section for a more complete account of these wild and wondrous goings-on.

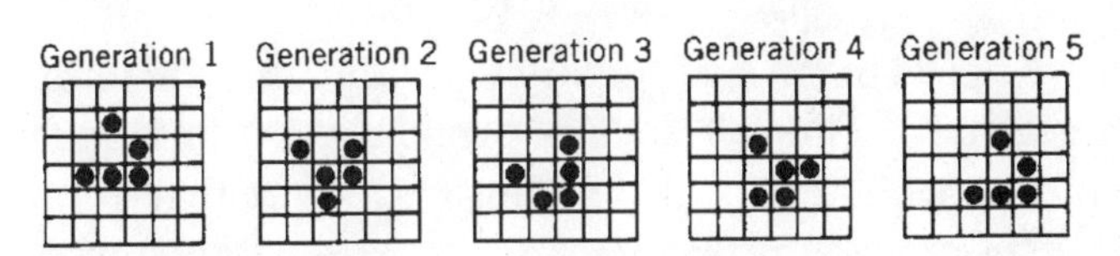

FIGURE 6.8. THE GLIDER

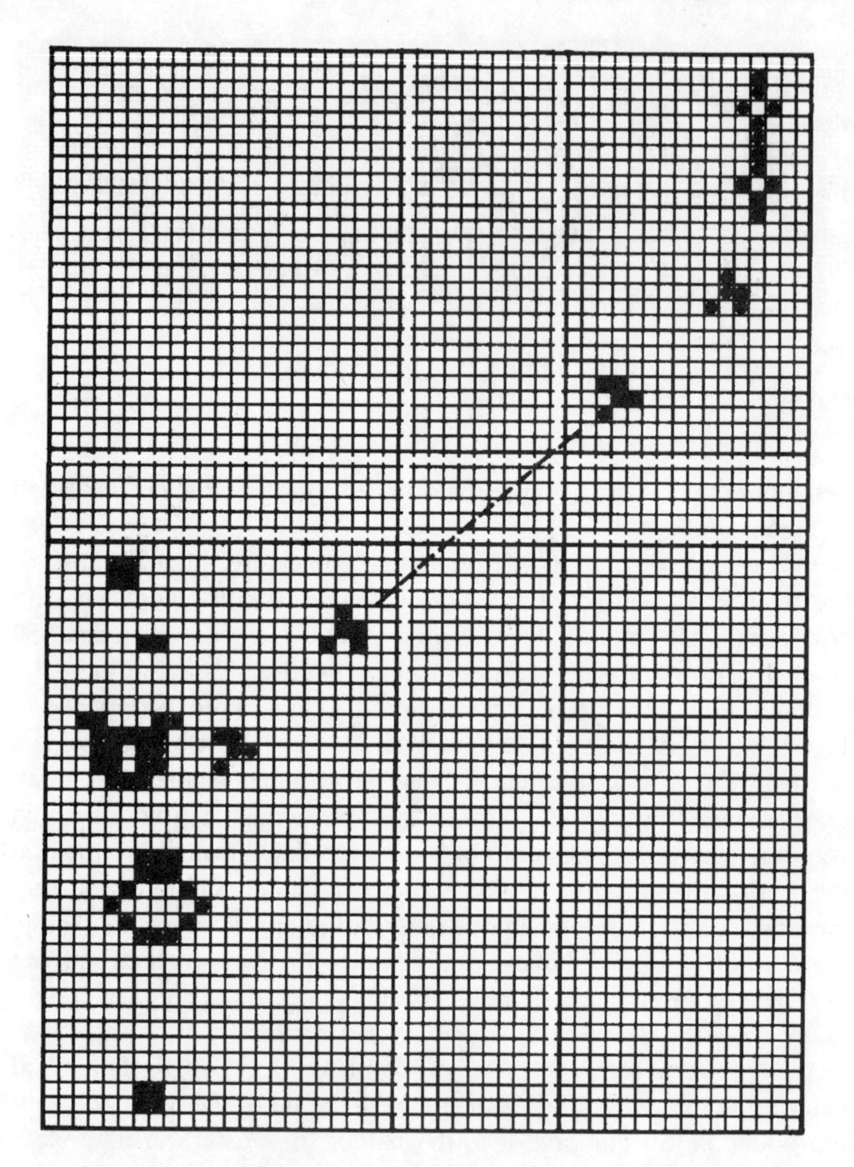

FIGURE 6.9. THE GLIDER GUN

Earlier we noted that it's customary to require that the reproduction of any genuinely self-reproducing configuration must be actively directed by the configuration itself. Thus we require that responsibility for reproduction reside *primarily* with the parent structure—but not *totally*. This means that the structure may take advantage of certain features of the transition rule's physics, but not to the extent that the structure is merely passively copied by mechanisms built into the transition rule alone. Von Neumann's requirement that the configuration make use of its stored information as both instructions to be interpreted and as data to be copied provides an appropriate criterion for distinguishing fake self-reproduction from the real thing. It's appropriate to close this discussion of emergence and self-reproduction by outlining the manner in which Conway showed that the *Life* CA rule admits configurations that are capable of self-reproduction in exactly this sense. With this bit of mathematical chicanery, Conway also showed that the simple *Life* rule is actually complicated enough to allow us to compute *any* quantity that can be computed. So it is another example of what in Chapter Four we called a universal Turing machine.

Conway's self-reproduction proof is based on the observation that Glider Guns (as well as many other *Life* objects) can be produced in Glider collisions. He then showed that large constellations of Glider Guns and Eaters can produce and manipulate Gliders to force them to collide in just the right way to form a copy of the original constellation. The proof begins not by considering reproduction per se but by showing how the *Life* rule allows one to construct a universal Turing machine. Since the *Life* universe consists of an array of ON-OFF squares on an essentially infinite checkerboard, what this amounts to showing is that one can construct a *Life* pattern that *acts* like a computer. This means that we start with a pattern representing the computer and a pattern representing its programming. The computer then calculates any desired quantity, which must then be able to be expressed as a *Life* pattern itself. For numerical computations this might involve having the *Life* computer emit the requisite number as patterns of a particular type or, perhaps, arranging to place the required number of patterns at some prespecified region of the display area.

The key to realizing the *Life* computer is the demonstration that any binary number can be represented by a Glider stream and that other *Life* patterns can be arranged to function as AND, OR and NOT gates, the building blocks needed for any computer. The biggest problem with these constructions is showing how the various streams representing the "wires" of the *Life* computer can be made to interpenetrate without losing their original structure. While wires and logic gates are all that's needed for any real-world computer, a universal computer needs something more: a potentially infinite memory! Conway's solution is to use the *Life* configuration termed a Block to serve as an external memory element. The Block consists of a 2 × 2 array of ON cells (see the fourth row of Figure 6.6) with the property that it is a so-called still life (i.e., it remains unchanged under the *Life* rule). The idea is to use the Block as a memory element outside the computer pattern and to use the distance of the Block from the computer to represent the number being stored. To make this scheme work, it's necessary to devise a procedure to move the Block, even though it's not in the computer. This can be accomplished by a tricky set of Glider-Block collisions. The end result of all these maneuvers is Conway's proof that the circuitry of any possible computer can be translated into an appropriate *Life* pattern consisting only of Guns, Gliders, Eaters and Blocks. But what about the other part of the self-reproduction process, the universal constructor?

The second part of the Conway proof is to show that any conceivable *Life* pattern can be obtained by crashing together streams of Gliders in just the right way. The crucial step in this demonstration is to show how it's possible to arrange to have Gliders converge from four directions at once in order to properly represent the circuits of the computer. The ingenious solution to this seemingly insoluble problem, termed *side tracking*, is much too complicated to describe here, but it provides the last step needed to complete Conway's translation of von Neumann's self-reproduction proof into the language of *Life*.

What would a self-reproducing *Life* pattern look like? For one thing, it would be BIG. Certainly it would be bigger than any computer or video screen in existence could possibly display. Moreover, it would consist mostly of empty space, since the design considerations require the use of extremely sparse Glider streams. The overall shape of the pattern could vary considerably depending upon design considerations. However, it would have to have an external projection representing the computer memory. This projection would be a set of Blocks residing at various distances outside the pattern's computer. Moreover, at least one of these Blocks would be special in that it would represent the blueprint of the self-reproducing pattern. (In actuality, the blueprint is the *number* represented by this Block.) For a detailed description of how the reproduction process works, the reader is referred to the literature cited in the To Dig Deeper section.

Now let me close this discussion of *Life* by noting the sobering estimate for how big such a self-reproducing *Life* pattern would have to be. Rough calculations suggest that such a pattern would probably require a grid of around 10^{14} cells. By comparison, a high-resolution graphics terminal for a home computer can display around 10^6 cells (pixels). To get some feel for the magnitude of this difference, to display a 10^{14} cell pattern, assuming that the pixels are 1 mm^2, would require a screen about 3 km (about 2 miles) across. This is an area about six times greater than Monaco! Thus, we can safely conclude that it is highly improbable that Conway's vision of living *Life* patterns will ever be realized on any real-world computer—even those likely to emerge from the rosiest of estimates of future computer technology. Too bad. It would have been fun to see these artificial lifeforms frolicking about in their checkerboard universe.

While we may not be able to display a living *Life* form, what our CA rules do often display is triangular patterns like those of the Type B attractor shown in Figure 6.4. Careful analysis shows that these triangles are formed when a long sequence of cells, which suddenly all assume the same value 1, is progressively reduced in size by local fluctuations induced by the particular CA rule. What is even more interesting is that if we keep track of the density of triangles having base length *n*, we find that for sufficiently large arrays this density is independent of both the initial configuration and the rule being used. In short, there seems to be a universal law governing the number and size of these triangles. This observation leads us to consideration of another kind of universal product of self-organizing systems.

THE MOST COMPLICATED THING IN THE WORLD

In 1733, Jonathan Swift commented on the infinitude of nature with the following bit of doggerel:

> So, Nat'ralists observe, a flea
> Hath smaller fleas that on him prey,
> And these have smaller fleas to bite 'em,
> And so proceed *ad infinitum*.

In a parody of this Swiftian verse, Lewis F. Richardson, the pioneering spirit behind today's computational approaches to weather forecasting, wrote of turbulent fluid flow that

Big whorls have little whorls
Which feed on their velocity,
And little whorls have lesser whorls,
And so on to viscosity.

These two light-hearted poems express another way that surprising behavior can occur as a consequence of the interaction among simple parts of a complex system. Rather than arising from mismatches in dimensional levels, as with the examples studied in the preceding chapter, these kinds of surprises come about because there is no obvious and natural scale against which to measure the behavior of many of the processes arising in the real world. What's observed on the largest scale of measurement repeats itself in an ever-decreasing cascade of activity at finer and finer resolutions. This sort of surprise arises in systems that are termed *self-similar*.

Experiencing the world ultimately comes down to the recognition of boundaries: self/non-self, before/after, inside/outside, subject/object and so forth. And so it is in mathematics as well, where we're continually called upon to make distinctions between categories—stable/unstable, computable/uncomputable, linear/nonlinear, real/complex—distinctions involving the identification of boundaries. In particular, in geometry we characterize the boundaries of especially important figures by giving them names like circles, triangles, squares and polygons. But when it comes to using boundaries to describe the natural world, these simple geometrical shapes fail completely. As IBM scientist Benoit Mandelbrot has expressed it, "Mountains are not cones, clouds are not spheres, and rivers are not straight lines."

To illustrate the point, Figure 6.10 shows two measurements of the coastline of Britain taken by Heinz-Otto Peitgen and his co-workers using measuring sticks of lengths 100 km and 50 km. Perhaps surprisingly, when we add up the length of Britain's coastline under these two different sets of measurements, we arrive at radically different estimates: about 3800 km when the scale unit is 100 km, nearly 6000 km when the scale is halved. Thus, as our measuring stick gets shorter, the length of the coastline seems to become longer! Intuitively we can understand this paradoxical result by recognizing that small bays of different sizes are overlooked when we use a ruler that's too long. But to get a better idea of what's going on, let's follow Peitgen's lead and do the same experiment for a circle.

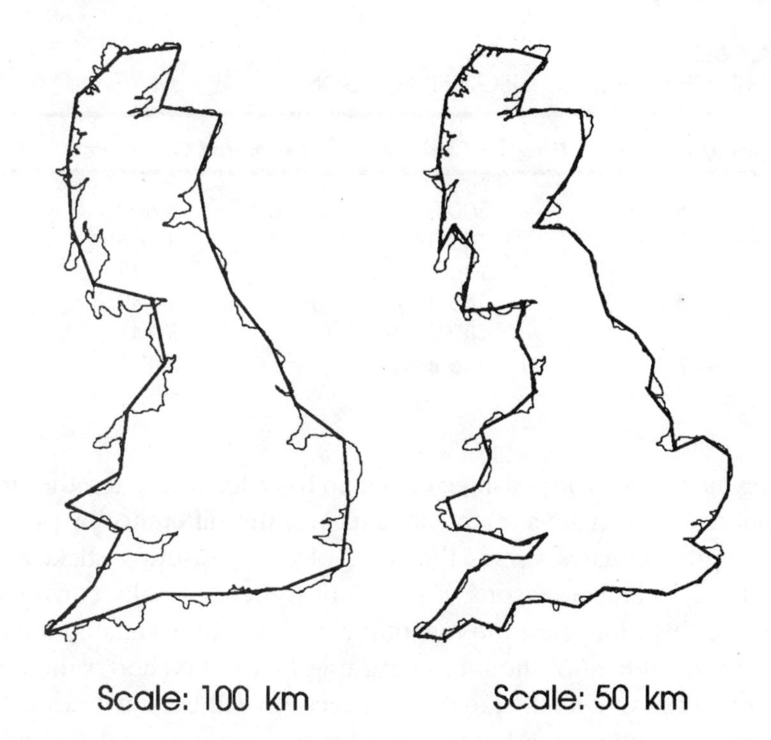

FIGURE 6.10. THE COASTLINE OF BRITAIN AT TWO DIFFERENT SCALES

Suppose we consider a circle of diameter 1000 km, so that its circumference is roughly the same size as the length of the coast of Britain. One way to estimate the circle's circumference is to inscribe a regular polygon (one whose sides are of equal length) within the circle and then add up the lengths of the sides of this polygon. By choosing polygons of greater and greater number of sides, this scheme, which can be traced to the ideas of Archimedes, leads to the estimates shown in Table 6.1. Surprise! In measuring the circle's length by using a measuring stick with finer and finer resolution (the length of the side of the inscribed polygon), the estimate does not go shooting off to infinity as it did with the coastline of Britain. Rather, it settles down to exactly what we expect it to be; namely, the exact circumference of the circle (which happens to be $2\pi \times 500$ km = 3141.59 km).

So what's the difference between these two experiments? Why did refining the unit of measurement lead to totally different results in one case while converging to the one true number in the other? Basically, the answer resides in the degree of "wiggliness" in the two curves

TABLE 6.1.
LENGTH OF A CIRCLE BY INSCRIBED POLYGONS

Sides of Polygon	*Length of Side (km)*	*Estimated Circle Length (km)*
6	500.0	2,600
12	258.8	3,106
24	130.5	3,133
48	65.4	3,139
96	32.7	3,141
192	16.4	3,141

being measured. The coastline of Britain has a lot of wiggles; the circle has none. We can get a quantitative feel for this difference by plotting the length estimates versus the scale of our measuring stick. For a variety of technical reasons, it turns out to be especially convenient to make this plot using the logarithms of both the estimated lengths and the *reciprocal* of the measuring unit (since the shorter the basic unit of measurement, the greater the precision in the estimated length). Figure 6.11 shows this log-log plot for the coastline and the circle. Here both the quantity u (the length estimate) and the number s (the scale of measurement) are taken in units of 1000 km. What we see from Figure 6.11 is that the regular curve (the circle) plots as a horizontal straight line, while the wiggly curve (the coastline) corresponds to a slanted straight line. From this, one might suspect that the greater the slope of the straight line, the more wiggly the curve. Figure 6.12, showing a similar calculation for several other geographic perimeters, lends support to this conjecture. And herein lies a tale.

The kind of curve describing the coastline of Britain is what in 1975 Benoit Mandelbrot christened a *fractal*. Fractals are curves that are irregular all over. Moreover, they have exactly the same degree of irregularity at all scales of measurement. So it doesn't matter whether you look at a fractal from far away or up close with a microscope—in either case you'll see exactly the same picture. If you start looking from a distance (i.e., with a "long" ruler), then as you get closer and closer (with shorter rulers) small pieces of the curve that looked like formless blobs earlier turn into recognizable objects, the shapes of which are the same as that of the overall object itself.

There are many examples of fractals in nature: ferns, clouds,

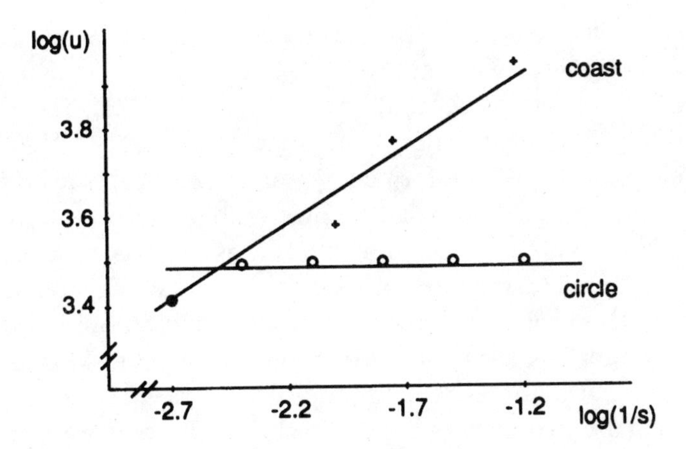

FIGURE 6.11. A LOG-LOG PLOT FOR THE LENGTHS OF THE COASTLINE OF BRITAIN AND A CIRCLE

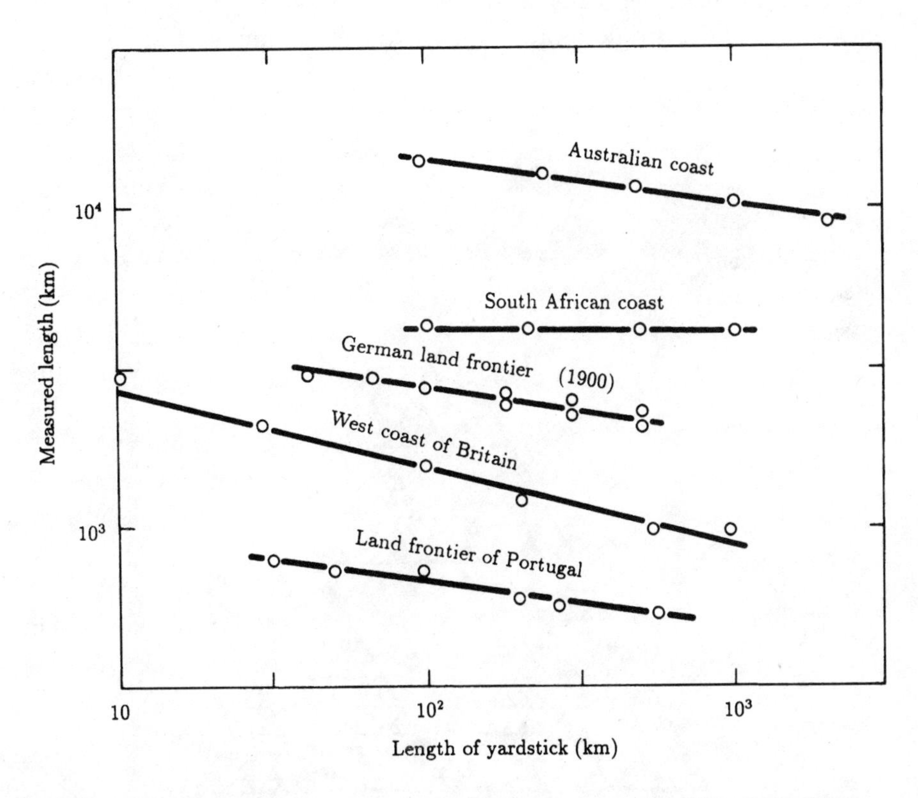

FIGURE 6.12. THE MEASURED LENGTH OF DIFFERENT GEOGRAPHIC PERIMETERS AT DIFFERENT SCALES

lightning bolts, coastlines, river basin networks and galaxies, to name but a few. We'll see some of these later. For now, let's look at a simple example of a fractal pattern called the *Sierpinski gasket*. The basic idea underlying the construction of this strange object is illustrated in Figure 6.13. The gasket is formed by starting with the black equilateral triangle at the top. The white triangle is then cut out of the center, leaving three smaller black equilateral triangles. This excision process is then repeated on the three black triangles, obtaining nine new black triangles. This process is then repeated indefinitely, doubling the detail at each stage. The final pattern is the famous Sierpinski gasket.

It turns out that there's a much more intuitively satisfying way of producing the Sierpinski gasket that enables us to see a most remarkable property of this figure. Think of a photocopy machine with a lens that reduces the original by one-half. Using this machine repeatedly to produce a sequence of images of the famous mathematician Karl Friedrich Gauss, we arrive at the picture shown in Figure 6.14. Now let's modify this single-reduction copy machine, giving it three lenses

FIGURE 6.13. THE SIERPINSKI GASKET

FIGURE 6.14. SINGLE-REDUCTION COPY MACHINE IMAGES OF GAUSS

instead of just one. The new machine is shown in Figure 6.15. Again we assume that each lens reduces the part of the image it scans by the same factor of one-half. Now we put in an image and ask what will emerge from the sequence of iterations as we run this feedback system. You might think that the images will get smaller and smaller—as with the images of Gauss—until they fade away into a single point. But this is not what happens at all! Figure 6.16 shows the end result of such an iteration starting with a single rectangle, where we color the reduced copies according to the respective lens system from which each copy is produced. The figure the three-lens machine generates is—the Sierpinski gasket.

It's tempting to think that the Sierpinski gasket emerges from the

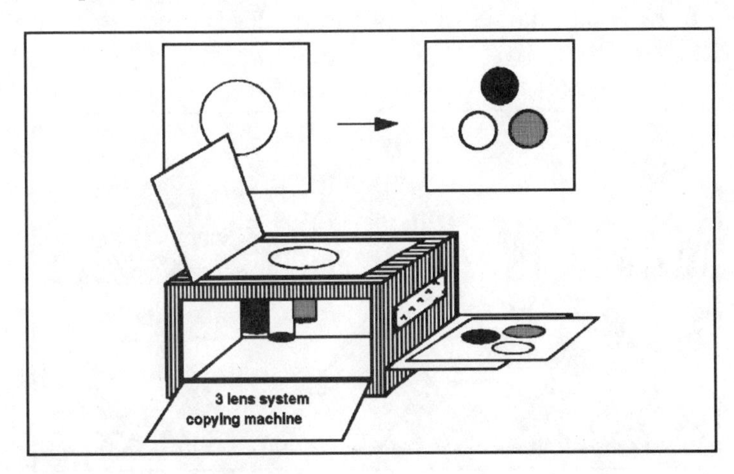

FIGURE 6.15. A MULTIPLE-REDUCTION COPY MACHINE WITH THREE LENSES

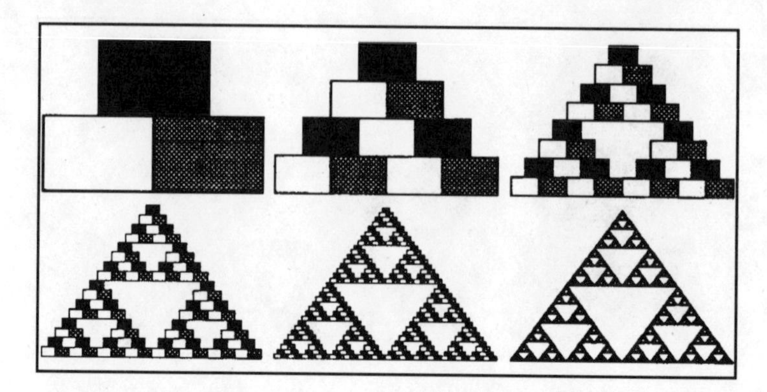

FIGURE 6.16. ITERATION OF A RECTANGLE WITH THE THREE-LENS COPY MACHINE

three-lens machine simply because we started our process with a rectangle. But this isn't the case. In fact, we can start with any image whatsoever and end up at the same place. To illustrate this crucial point, consider starting with the letters NCTM. The successive iterates of this starting image are shown in Figure 6.17. Again we end up with the Sierpinski gasket. This suggests that there is something universal about the Sierpinski gasket. Unfortunately, this is not quite the situation. But what is true is that a closely related object, the *Sierpinski carpet*, does have the remarkable universality property that any one-dimensional object whatsoever can be hidden somewhere in the carpet. The construction of the Sierpinski carpet is shown in Figure 6.18. For more information on this point, and for a wealth of additional detail on these fascinating objects, the reader is urged to consult the material cited in the To Dig Deeper section.

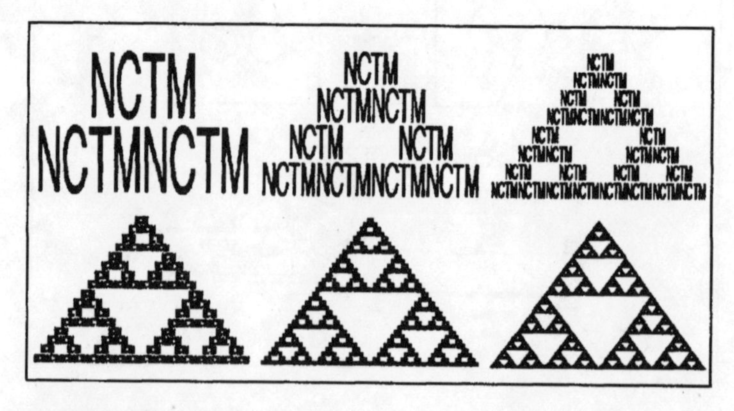

FIGURE 6.17. ITERATION OF THE IMAGE NCTM WITH THE THREE-LENS COPY MACHINE

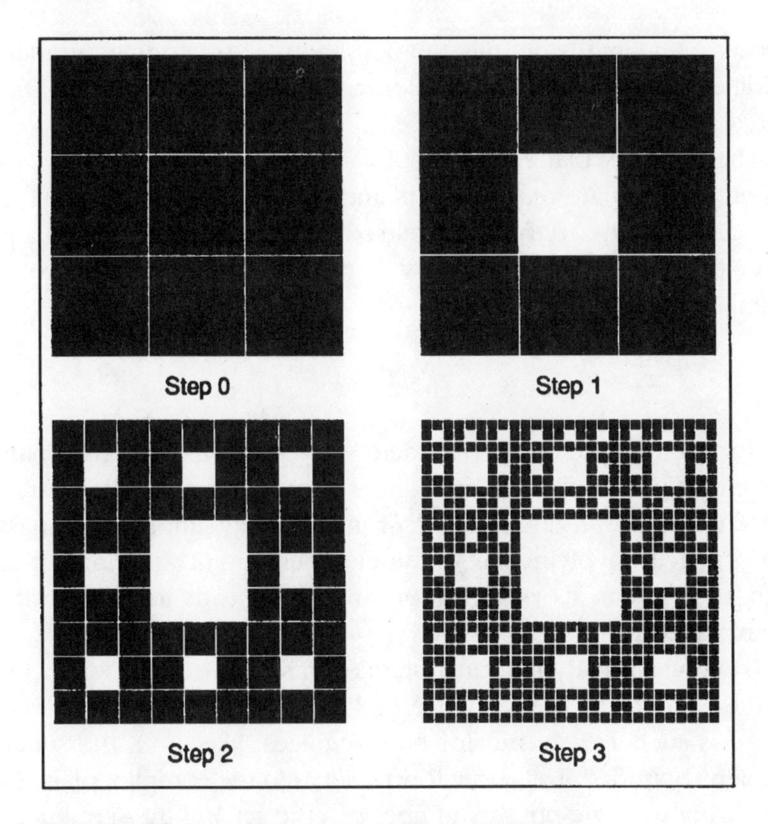

FIGURE 6.18. CONSTRUCTION OF THE SIERPINSKI CARPET

The Sierpinski gasket and carpet display what we call *linear self-similarity*, in the sense that any part of the object is exactly like the whole. But the most important fractals are not this rigidly self-similar, including the most famous fractal of all, the celebrated *Mandelbrot set*. No account of fractals can avoid discussion of this "most complicated object in the world."

Mandelbrot, Julia and Domains of Attraction

We have seen in earlier chapters that the most important thing one can know about a dynamical system is its attractors. Associated with each attractor is a set of initial states whose trajectories end up in that attractor under the action of the dynamical system's vector field. These starting points are called the *domain of attraction* of the attractor. In the Beer Game example of Chapter Three, we saw that the boundary of the domain of attraction can be a very complicated curve or surface. Here we'll see that even for the simplest kind of nonlinear dynamical process, one involving only the operations of squaring a number and addition, the boundary of the domain of attraction can be as complicated as it's possible to be.

The complex plane consists of all numbers of the form $z = a + bi$, where a and b are real numbers and i is the square root of -1. Here a is called the *real part* of z, while b is the *imaginary part*. Fix a point c in the complex plane. Now construct a sequence of complex numbers by the rule

$$z_k = z_{k-1}^2 + c, \quad z_0 = 0 \quad k = 1, 2, \ldots \qquad (*)$$

So for each choice of the parameter c, we have a different dynamical system. If you happen to be wondering why we chose this particular system, the reasons are twofold: (1) it's about as simple as a dynamical process gets, involving just the single operation of squaring a number and adding it to its predecessor, and (2) it leads to interesting and unexpected results.

Looking at the dynamical system (*), it's clear that for some choices of the parameter c the sequence $\{z_k\}$ will diverge to infinity. The point $c = 2$ is such a case. But for other choices, like $c = i$, the sequence remains bounded. Let's now form a set M in the complex plane in the following way: we put the number c in the set M if the sequence $\{z_k\}$ does *not* go off to infinity for that value of c; otherwise, c does not belong to the set M. So, for instance, the point $z = 2$ is not in M, while the point $z = i$ does belong to it. The set M, called the *Mandelbrot set*, is shown in Figure 6.19.

It has been shown that if the starting point z_0 lies in the interior of the Mandelbrot set, the dynamics (*) generates a trajectory that is orderly and well-behaved for any value of c. But if z_0 lies outside M, the trajectory—while perfectly deterministic—is wild and disorderly, wandering all over the complex plane. The boundary of M, which separates the orderly and disorderly behaviors, turns out to be unbelievably messy. As one might surmise from inspecting Figure 6.19, new copies of the entire Mandelbrot set continue to "bud off" from the original at all scales of observation. Furthermore, there are many additional structures that appear at various magnification levels. To illustrate, Figure 6.20 shows a magnified view of part of the boundary of the set M. New structures continue to appear as one goes to higher and higher magnifications, the set M remaining forever intricate on

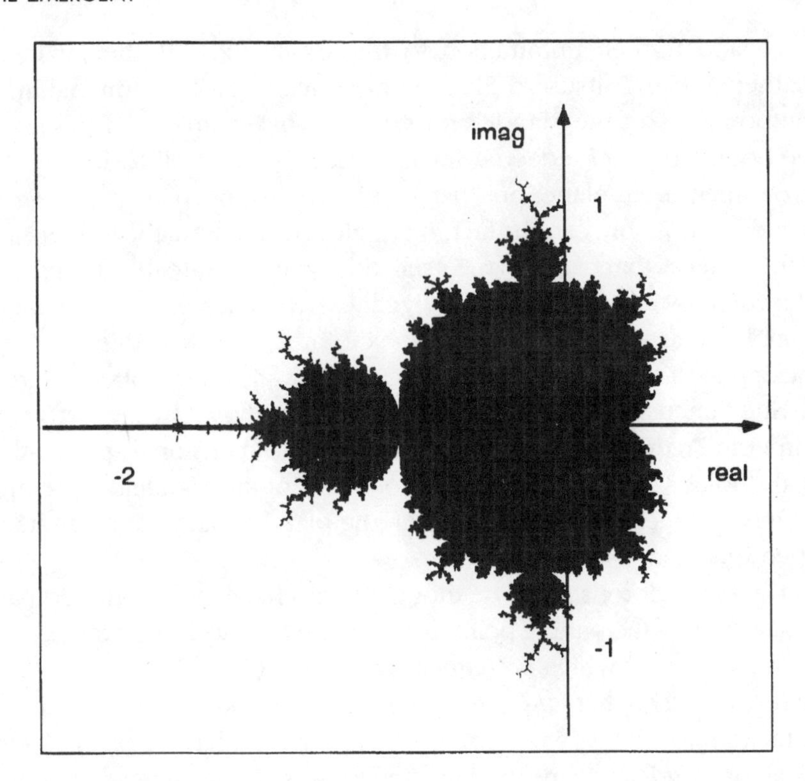

FIGURE 6.19. THE MANDELBROT SET

any scale. It's this kind of never-ending sequence of new patterns that leads many to claim that the Mandelbrot set is the most complicated object known to humankind.

FIGURE 6.20.
MAGNIFIED VIEW OF PART OF THE BOUNDARY OF THE MANDELBROT SET

To add further ammunition to this claim, in 1991 the Japanese mathematician Mitsuhiro Shishikura proved that the dimension of the boundary of the Mandelbrot set is 2. This means that not only is the boundary of M indeed a fractal, since the set itself resides in the two-dimensional plane, but that it "wiggles" as much as any curve in the plane possibly can. In fact, it wiggles so much that its dimension is the same as that of the plane itself. So if you measure the complexity of a curve by its irregularity, or wiggliness, then no curve can be more complex than the boundary of M. Note, however, that this does *not* mean that the boundary of M is a curve that fills up the plane. Merely having dimension 2 is not enough for this, as is shown by the Sierpinski carpet in Figure 6.18, which is a kind of two-dimensional sponge that is all "holes." Whether or not the boundary of the Mandelbrot set fills up a two-dimensional continuum in the plane remains, at present, an open question.

The Mandelbrot set can be thought of as characterizing the domain of attraction of the starting point $z_0 = 0$ for the *family* of maps $z \to z^2 + c$, each member of which is "named" by a particular value of c. But this is not usually what we mean when we speak of the domain of attraction. Normally, we consider a *single* system having one or more attractors, and we want to characterize those initial states leading to the different attractors. This was the situation, for example, with the chaotic examples discussed in Chapter Three. Just as with those examples, if a system has a strange attractor, the boundary of that attractor has remarkable fractal properties.

To see this, let's go back to the Mandelbrot mapping considered earlier:

$$z_k = z_{k-1}^2 + c, \quad k = 1, 2, \ldots$$

We now fix the parameter value at $c = -1$ and look at the set of initial points z_0 leading to trajectories $\{z_k\}$ that remain finite as k gets larger and larger. The boundary of the black part of Figure 6.21 shows these points, while the black interior is the strange attractor for this system. The set of boundary points is called the *Julia set* of the system, while the filled-in black part is called the *filled-in Julia set*. For the sake of completeness, let's note that the complement of the Julia set (the white part) is called the *Fatou set*. These names owe their origin to the French mathematicians Gaston Julia and Pierre Fatou, who studied this kind of set in the early part of this century.

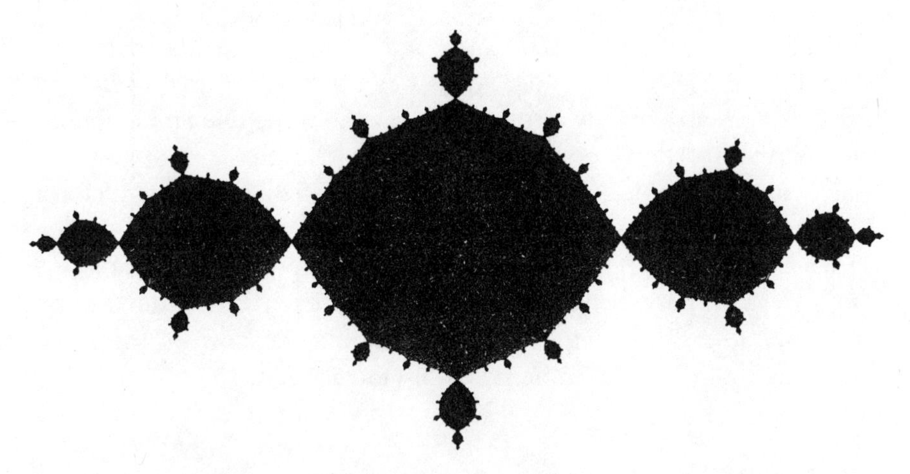

FIGURE 6.21. THE JULIA SET OF THE MAP $z \rightarrow z^2 - 1$

It's interesting to see what happens to the Julia set for the system (∗) when $c = 0$. In that case, the iterates are just the powers z, z^2, z^4, z^8 and so on. So if the initial point z_0 lies inside the circle in the complex plane centered at the origin and having radius 1, then its trajectory remains bounded. But if the starting point is outside this unit circle, the trajectory flies off to infinity. Finally, if the starting point lies on the boundary of the circle, the successive points move around on the circle indefinitely. Therefore, for this system the Julia set is simply the unit circle. But amazing things start to happen to the Julia set when we let c move away from zero.

For even the smallest nonzero value of c, the Julia set gets distorted—but not into a smooth, approximately circular curve. Instead the curve becomes a fractal. Figure 6.22 shows a sequence of distortions of the original unit circle as we gradually increase the magnitude of c, going from the "almost circle" in (a) to the so-called Douady rabbit in (b) and on to a set of disconnected islands in (c).

So the rule (∗), simple as it is, shows us that the boundary of the

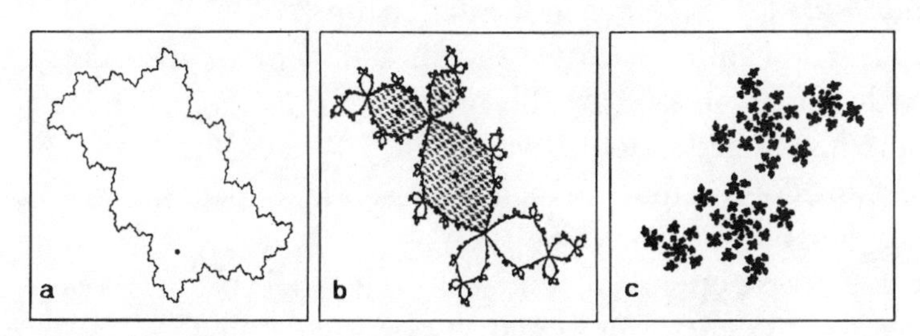

FIGURE 6.22. THE JULIA SET FOR VARIOUS NONZERO VALUES OF c

domain of attraction of a strange attractor can be just about as complicated as a mathematical object can get. But to get a real feel for the depth and beauty of these objects, the more detailed literature cited in the To Dig Deeper section is must reading.

Fascinating as things like the Sierpinski gasket and the Mandelbrot set are as objets d'art, it's at least defensible to ask about the degree to which fractals are anything beyond mere mathematical curiosities. Do they have any connection to everyday events? We devote the concluding sections of this chapter to a tour of some of the more exotic fauna frolicking in the fields of Fractal Land.

FROM BACH TO ROCK AND BACH AGAIN

Legend has it that in 1735 Swiss physicist and mathematician Daniel Bernoulli was thrown out of the house by his father, mathematician Johann Bernoulli, when the Paris Academy awarded them jointly a prize for work on planetary orbits that the elder Bernoulli felt should have been his alone. Perhaps this bit of patriarchal pique was the spark that touched off Daniel's interest a couple of years later in what economists now call *utility theory*. In any case, by 1738 Bernoulli *fils* had devised a quantity called a utility function, measuring an individual's social well-being, which he used to characterize people's behavior.

The basic idea underlying Bernoulli's theory is rather simple. He argued that the greater someone's level of income, the less important is any particular income shift. If we denote the income level by f and any change in income by Δf, Bernoulli's claim is that shifts in income for two individuals at different income levels are equivalent if the ratio $\Delta f/f$ is the same for both. In other words, if I enter into a transaction with you that involves each of us committing half our income, we would both get the same utility from the transaction—regardless of the *absolute* levels of our individual incomes.

Bernoulli went on to argue that the utility of a transaction f for an individual is expressed by the rule $U(f) = \log(f/f_0)$, where f_0 is the income level needed to just sustain life. What's important about this logarithmic form is that the logarithm is independent of the units (e.g., dollars, lira, horses, ounces of gold) used to measure the income level f and does not depend on the scale (i.e., the absolute level) of the process. Sad to say for Bernoulli, it seems that human behavior cannot be explained by such a simple principle as maximizing utility. Nevertheless, the fact that both the logarithm and the relative change in a quantity f are independent of the scale used to measure the process leads to a phenomena termed *1/f (one over eff) behavior*, which *does* occur regularly in other areas of natural and human activity.

A good example of 1/f behavior arises in linguistics. Suppose we list the words in a particular language in decreasing order of how often they occur in a broad sample of text taken from that language. The *rank* of a word is simply its position on this list. So, by definition, the rth word on the list has rank r. Around 1950, George Zipf discovered what's now termed *Zipf's Law*, which gives a quantitative description of how the relative word frequency is related to the word rank. Symbolically, we can write Zipf's Law as $f(r) \approx 1/(r \log(1.78R))$, where R is the number of words in the language involved. For example, a reasonably comprehensive dictionary of the English language lists about $R = 12{,}000$ words. Thus, by Zipf's Law the relative frequencies of the highest-ranking words, *the, of, and* and *to*, should be 0.1, 0.05, 0.033 and 0.025, respectively. The logarithmic form of Zipf's Law makes it clear that it has the 1/f form noted above.

If we graph the frequency of a word's appearance versus its rank in a large block of typical English text, measuring both on a logarithmic scale, we obtain the diagram shown in Figure 6.23. A perfect 1/f law would have a slope exactly equal to –1, which is shown as the dark line on the graph. The relative frequencies of nine representative English words are indicated by the arrows, showing the excellent empirical agreement with the 1/f rule.

Interestingly enough, studies have shown that a monkey hitting the keys of a typewriter randomly will also generate a "language" that obeys Zipf's Law. If we plot the words of this monkey language on a graph similar to that for English, we find the slope of the line equals –1.068, only slightly different from the slope of a perfect 1/f phenomenon. Manfred Schroeder has calculated that with a nine-letter alphabet and a probability of hitting the space bar of 0.1, the *median* word rank of the monkey language is an astronomical 1,895,761. This means that if you take a word at random from a monkey text, you would need a list containing this many words before there was a 50–50 chance of finding the randomly selected word on your list. By way of contrast, the median word rank of English texts is around 100 for a

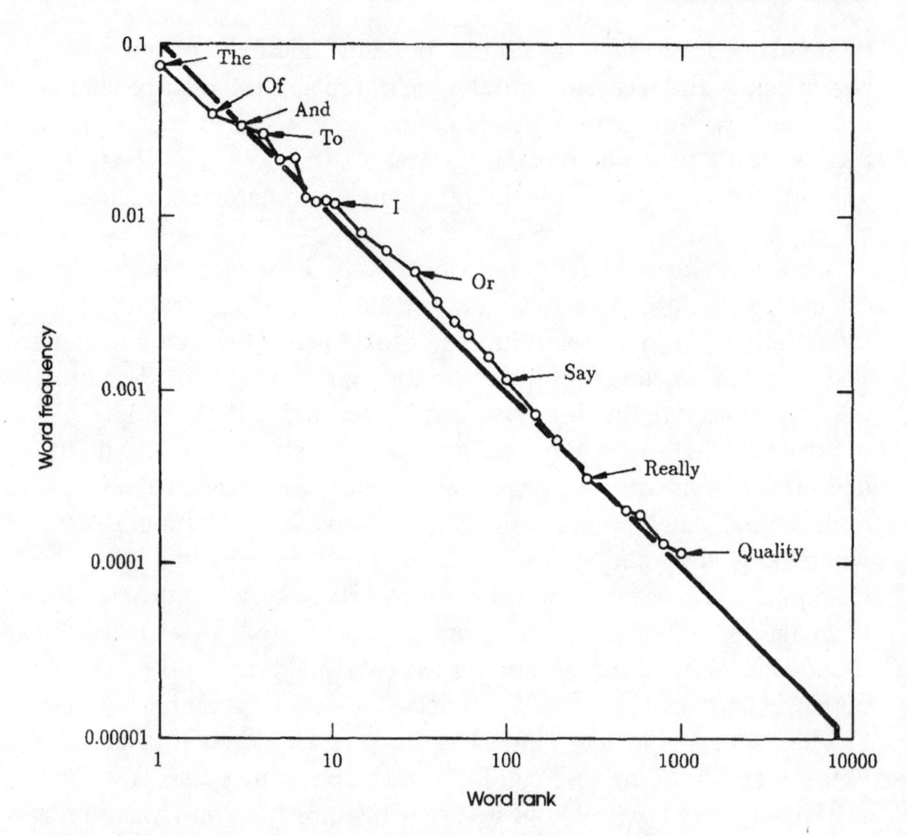

FIGURE 6.23. WORD FREQUENCY VERSUS WORD RANK FOR ENGLISH

typical newspaper article and ranges up to 500 for very literate writers. Thus, the monkey language, while adhering closely to Zipf's Law, is a very wordy language indeed.

We have seen that in describing a dynamical process, the various time scales (frequencies) contributing to the process run across a spectrum characteristic of the process. This spectrum shows how the energy of the process is distributed among different types of motion. For instance, a simple vibrating spring has a spectrum consisting of a single frequency. But a purely random series of numbers, coming from, say, the movement of a gas molecule in a container, has a very broad spectrum since many frequencies contribute to its underlying dynamics. A fractal process, on the other hand, has no characteristic frequency or scale, and its frequencies form what's called an *inverse power spectrum*. This is usually expressed as $1/f^{\alpha}$, where f is a frequency and α is some positive number. When $\alpha = 1$, we obtain the 1/f spectrum just considered.

At this juncture we might well inquire, What's the connection between a fractal and a phenomenon like language that obeys a 1/f rule? For the answer, it's instructive to talk about another kind of language—the language of music.

White, Brown and Pink Music

Suppose you take the score of a piece of music and calculate the power spectrum of the relative frequency intervals x between successive notes. Let's call this function $p(x)$. Carrying out this exercise for Bach's *First Brandenburg Concerto*, you'll find that over a large range of intervals the spectrum has the form $p(x) = c/x$, where c is a constant that characterizes this particular piece of music. Thus, Bach's music would be described as sound of the type communication engineers term *1/f noise*. (Note: Following the custom in certain engineering circles, we are using the term *noise* as being synonymous with *sound*.)

This very specific type of functional relationship between the frequency and the intervals between successive notes shows up in the spectrum of amplitudes, too, as indicated in Figure 6.24 for Bach's *First Brandenburg Concerto*. The question, of course, is why Bach and so many other composers create music that seems to obey this 1/f rule.

Part of the answer to this puzzle lies in the observation that for any piece of music to be "interesting," it should be neither too regular (like most modern rock and C&W tunes), nor too unpredictable (like many avant garde compositions).

To see what's involved in satisfying these conflicting conditions, suppose you put a tape recording of a piece of music into a recorder and play it at double the normal speed. For most types of music, what results is a highly distorted, squawky output from the speakers that sounds nothing at all like music. But for some special types of sounds, speeding up the recorder has no effect at all on the output. A good example is purely random noise, in which every frequency occurs with equal likelihood at any given moment. In the case of this so-called white noise, all you hear is a steady hiss regardless of the recorder's speed. We could also try putting what's called *Brownian noise* into the recorder. This is a kind of sound in which there is an equal chance of every frequency *difference* appearing in successive time intervals. With this type of noise, there is a strong correlation between the frequency at one moment and the next, but the frequency differences

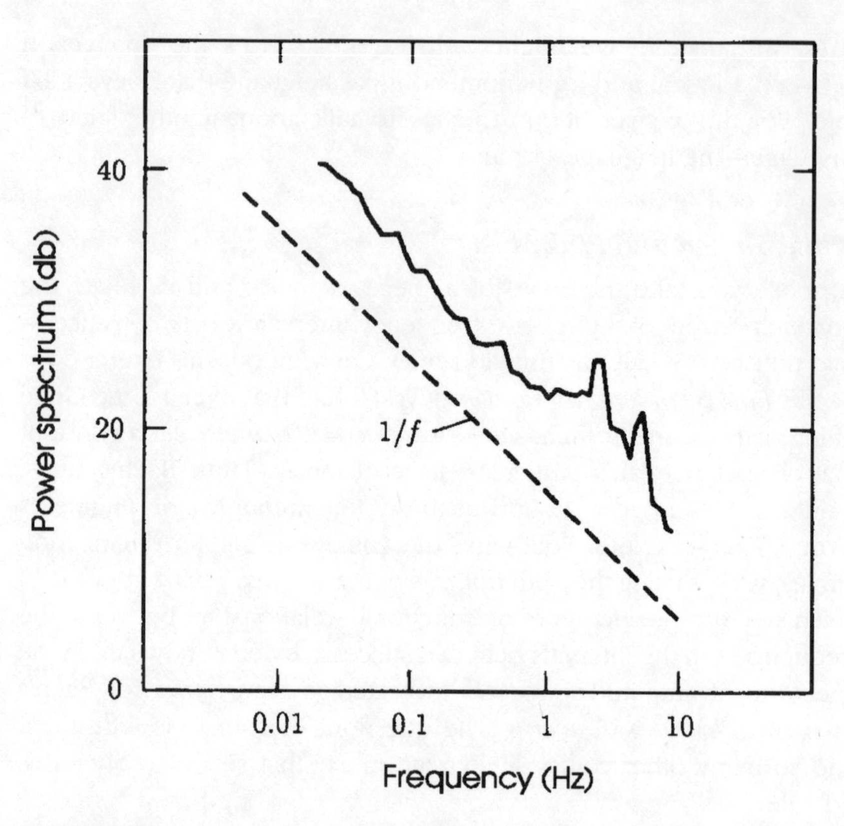

FIGURE 6.24. AMPLITUDE SPECTRUM FOR THE *FIRST BRANDENBURG CONCERTO*

are independent. As with white noise, the output from the recorder also sounds the same with brown noise regardless of how fast the tape is played.

As nature would have it, this kind of *output invariance* is characteristic of all processes obeying the inverse power-law rule outlined above. As noted above, these processes are not characterized by any natural frequency or scale; they look (and sound) the same on all time scales. So when it comes to pleasing music, the relationship linking the intervals between successive notes and how often these intervals occur should not behave like the highly correlated, monotonically sounding "brown" noise. Nor should the music sound like completely unpredictable "white" noise. Just as with Goldilocks's porridge, good music should be not too hot and not too cold—but just right. This suggests looking at the intermediate "pink" noise case, which happens to correspond to a 1/f power spectrum relationship. (By all rights this should probably be termed "tan" noise since it's midway between brown and white. But engineers have their own brand of logic, and the term pink more accurately matches the color spectrum of this kind of process than does tan.)

Figure 6.25 shows examples of music composed according to each of these frequency patterns. Part (a) shows white music, produced from independent notes (i.e., music in which there is no correlation, or "memory," between a note and its predecessor). Part (b) is brown music, composed of notes with independent frequency increments. Part (c) consists of pink music, in which there is a weak correlation between the frequency and duration of successive notes. This last kind of music is 1/f noise.

Now let's make the connection between music and fractals. When you look at the score of a piece of music, it's vaguely reminiscent of looking at the profile of a mountain range or, perhaps, the skyline of a large metropolis like New York or Chicago. This kind of profile is

FIGURE 6.25. (A) WHITE, (B) BROWN AND (C) PINK MUSIC

what we've termed earlier a *fractal*. So we could compose "mountain music" by just assigning notes on the basis of how the profile moves up and down. The composer Villa-Lobos actually carried out this experiment using the mountain ranges near Rio de Janeiro. But most such music is too "brown" to be aesthetically interesting. In other words, most mountain ranges have profiles that are too highly correlated. But the 1/f rule seems to govern many other natural phenomena, including those phenomena that appear to have artistic merit. So the connection between 1/f processes and fractals is that the profile of naturally occurring 1/f phenomena like some mountain ranges and coastlines turns out to be a fractal. Now let's get back to the music.

The similarity laws obeyed by fractal objects suggests a way of "compressing" the music of a Bach or a Mozart to its irreducible essence. So just as we can recognize Bob Hope by his ski-jump nose or Alfred Hitchcock by his portly profile without having to see either of these eminent personages in any greater detail, we would also like to be able to single out the music of a particular composer from a similar kind of skeletal outline.

This notion suggests the possibility of using a distilled Bach "essence" to compose new Bach-like music, a challenge that was taken up recently by Swiss physicist Kenneth Hsu, using the frequency and amplitude spectrums of Bach's music to determine each note's relationship to its neighbors. By removing notes from several of Bach's inventions, Hsu found that basic patterns persisted in the fractal reductions of the music, even if what remained contained as little as one sixty-fourth of the original notes. Thus, music recognizable as Bach survives fractal reduction. In fact, to some ears this "reduced Bach" gives the impression of an economy of frills and ornamentation. For those of a musical orientation, Figure 6.26 shows one of these reductions in the case of Bach's *Invention 5*. With the happy thought of an almost inexhaustible supply of new Bach-like music emerging from the computer laboratories of the future, let's now leave the rarefied heights of these humane arts and consider the appearance of fractal phenomena in other areas of human concern. It's hard to think of an area more people have spent more time worrying about than the world of easy money. In particular, our next story deals with the surprising appearance of fractals and self-similarity in answering the eternal question for casino gamblers everywhere, How much should I bet?

FIGURE 6.26. BACH'S *INVENTION 5* IN ITS ORIGINAL AND FRACTAL FORMS

CLIMBING THE DEVIL'S STAIRCASE

Nick "The Greek" Dandolos (1883?–1966) was generally regarded as the most flamboyant—if not famous—gambler of his era. He reputedly won or lost over $500 million, beginning in 1919 when he hit the bank at Monte Carlo three times for $20,000. The range of his friendships included Albert Einstein, the Prince of Wales (later Edward VIII) and Jack Dempsey. And when it came to homespun philosophy, The Greek made it into the history books with his remark that "The only difference between a winner and a loser is character." In conjunction with this pretty banal observation (after all, character is about the only important difference you can find between people anyway), The Greek also had a practical prescription for winning at the tables. His version of fortune's formula is summed up in the statement

> Remember that old percentage is always back there, grinding away. The only way you can keep it from slowly grinding your bankroll to a pulp is to win as much as possible in as few bets as possible.

Interestingly enough, modern mathematics and the theory of fractals provide some ammunition backing up Nick the Greek's claim.

Suppose we place bets on either black or red turning up on successive spins of a roulette wheel. Then our probability of success on any single spin of an American wheel with its twin zero slots is $p = 18/38$ (European roulette wheels have only a single zero). Moreover, let's assume we start with \$100, hoping to run it up to \$20,000 before going broke. Classical arguments from probability theory show that if we follow a strategy of *timid play*, betting a measly *fixed* amount of \$1 on each play, the likelihood of attaining our goal before being busted by the casino is about 3 chances in 10^{911}—indistinguishably close to impossible. Now suppose we decide to wager a fixed amount of \$10 per play. In this case, the chances of winning twenty grand before going broke improve to about one in 10^{91}, an enormous relative improvement but still negligible in absolute terms. Nevertheless, the message is clear: if at each play of the game you have less than a 50–50 chance of winning, larger bets improve the chances of reaching a set goal before going broke. This leads to the strategy of *bold play*.

Following the dictates of bold play, on each turn of the wheel we bet our entire bankroll if that amount is not greater than half the goal (in this case \$20,000). Otherwise, we bet the difference between the goal and the amount of money we currently have in hand. It can be shown that this strategy maximizes the chances of reaching the goal in a fair game (one in which there is a 50 percent chance of success on each trial). In roulette, of course, the chances are not equally balanced in this way, since the likelihood of success on any trial is $18/38$, which is less than 0.5. For the specific case we're looking at here, starting with \$100 and following the strategy of bold play, there are three chances in a thousand of reaching the \$20,000 plateau before going broke—about ninety orders of magnitude better than the chances when using timid play.

To analyze bold play, let's normalize things so that our bankroll ranges between 0 and 1, the goal being to reach 1 before going bust. Let $m(x)$ be the likelihood of success when starting with an initial amount x, and let p represent the probability of winning on any single play of the game. Bold play says that if the initial bankroll, x, is less than 0.5, you should bet everything you have. But if our starting bankroll x is greater than or equal to 0.5, we should bet $1 - x$. In the first case, if we're to reach our goal of 1, we have to win on the first play (with probability p) and then, starting from the new bankroll of $x + x = 2x$, we must go on to eventual success (which, by definition, has probability $m(2x)$). Thus, the product $p \times m(2x)$ is the probability of winning on the first play *and* going on to reach the goal. A similar, but slightly more involved, line of reasoning leads to the quantity $p + (1 - p) \times m(2x - 1)$ for the likelihood of reaching the goal when starting with a bankroll x that is greater than 0.5. Symbolically, we can summarize the situation thus far in the following pair of equations:

$$m(x) = \begin{cases} pm(2x) & \text{for } 0 \le x \le 0.5 \\ p + (1 - p)m(2x - 1) & \text{for } 0.5 \le x \le 1 \end{cases}$$

The function $m(x)$ is shown in Figure 6.27 for a situation in which the probability of success on any single play of the game is $p = 0.25$. Looking carefully at Figure 6.27, we note the following two important properties of this function:

1. The function $m(x)$ increases steadily with x. So the greater our initial bankroll, the better our chances of reaching the goal. While this is fairly obvious on the basis of in-the-casino empirical results, it's reassuring to see the mathematics reflecting this commonsense reality.
2. The probability of success increases only on a set of capital levels x having zero measure in the interval between 0 and 1. This means that if the probability function $m(x)$ were to be magnified, we would see it as composed of many small "stair steps." And the only places where the function increases is at those points where we move from one step to the next. It can be shown, however, that this set of points has a negligible (actually, zero) measure in the unit interval. Such a curve is what's called a *devil's staircase*. Thus, it's only at a mathematically negligible set of bankroll levels that the chances of success can be improved by increasing the starting bankroll.

OK, so what does the function $m(x)$ have to do with self-similarity? Let's assume that the bankroll, x, is less than 0.5, so we're in the first case discussed above. Then the connection with self-similarity comes in by noticing that if we divide x by 2 and multiply $m(x)$ by p, we reproduce the left half of Figure 6.27. So the probability function $m(x)$ is what's called *self-affine*. And it is exactly this property that gives rise to the fractal-like devil's staircase type of structure just noted. But the flow of money in the casino is nothing compared to what moves

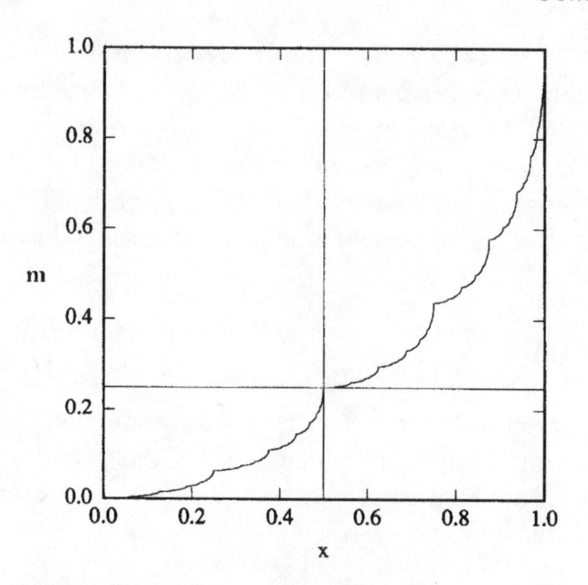

FIGURE 6.27. THE FUNCTION $m(x)$ FOR $p = 0.25$

about the world's financial systems each day. So let's leave the felt-top jungles and cast our gaze in the direction of Wall Street.

And So it Flows

Unlike money, water has a tendency to flow from where it's abundant to where it isn't. But just like the ebb and flow of prices on speculative markets or the fluctuations of a gambler's bankroll, water also displays fractal behavior in its multifarious meanderings. So in preparation for tackling the problem of stock price movements and the fluctuation of currency exchange rates, let's look at how hydrologists have grappled with the question of measuring the fractal nature of the ups and downs of water reservoir levels.

Suppose we have a reservoir that's fed by the discharge of water from a lake. In any given year the reservoir receives an influx from the lake and a regulated volume of water is released. If we let $x(t)$ denote the inflow in year t, then the average inflow over a period of T years is simply $(1/T)[x(1) + x(2) + \cdots + x(T)]$. Call this average inflow $\bar{x}_T$. Now define the departure of the inflow from this average over a t-year time horizon to be $X(t, T) = [(x(1) - \bar{x}_T) + (x(2) - \bar{x}_T) + \cdots + (x(t) - \bar{x}_T)]$. The difference between the minimum and the maximum accumulated inflow over a period of T years is what we call the *range*,

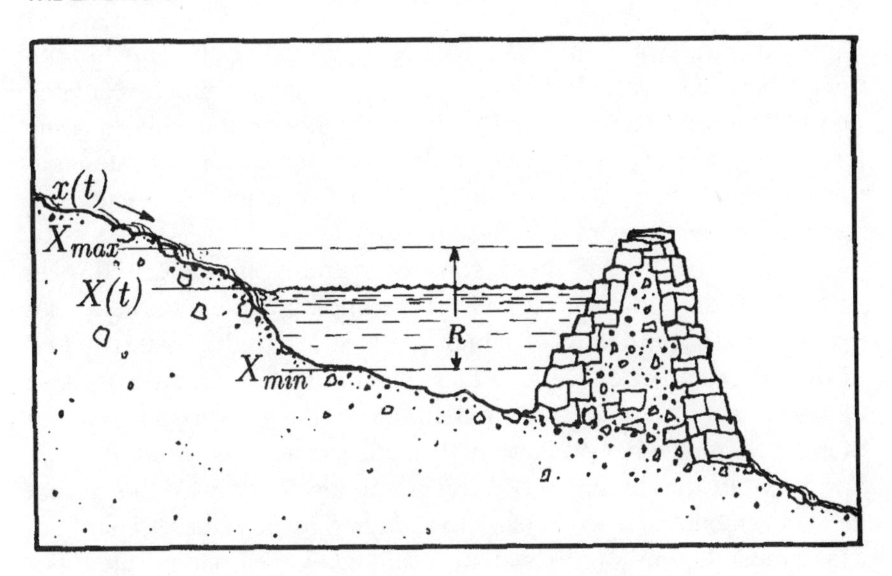

FIGURE 6.28. A WATER RESERVOIR WITH INFLOW $x(t)$ AND RANGE R_T

denoted by R_T. This overall situation is shown schematically in Figure 6.28.

It's clear, I think, that the range depends on the time period T and that it can only increase with longer and longer time periods T. In the late 1940s, hydrologist H. E. Hurst investigated many natural phenomena, like river discharges and tree ring growth, using the range R_T and the sample standard deviation

$$S_T = \{(1/T)\,[(x(1) - \bar{x}_T)^2 + (x(2) - \bar{x}_T)^2 + \cdots + (x(T) - \bar{x}_T)^2]\}^{1/2}$$

What Hurst discovered is that a large number of natural processes seem to be governed by a simple relation involving what's now called the *rescaled range*, defined by the dimensionless ratio R_T/S_T. This quantity, in effect, scales the range by taking the standard deviation as the unit of measurement.

Hurst's results showed that in a great many phenomena, ranging from flood levels on the Nile to trends in global temperature, changes seem to obey the empirical relation $R_T/S_T = T^H$, where H is a number now called the *Hurst exponent*. It can be shown that when the individual measurements $x(t)$ are independent from one time period to the next, the Hurst exponent must tend to 0.5 for a large enough time horizon T. What's remarkable about Hurst's work is that an impressive array of empirical evidence suggests that the Hurst exponent differs substantially from 0.5 for many natural processes. So if you believe the data, it's hard to reject the notion that there's some long-term "memory effect" present in processes like sunspot fluctuations, river discharges and rainfall levels, all of which have values of H differing greatly from 0.5.

To understand *why* these sorts of natural phenomena display values of H differing from 0.5, Hurst dreamed up an experiment with playing cards, the details of which are best left to the material cited in the To Dig Deeper section. This experiment suggested that these kinds of natural processes "remember" what's happened earlier, thereby creating what probability theorists call a *biased random walk*. For example, the discharge of a river depends not only on the current level of rainfall but also on earlier rainfalls. Similarly, the discharge of a lake must depend on the water present in a large drainage area. But the amount of water in the area will increase in periods of heavy rainfall, with the excess amount of water then being stored to contribute to the discharge in drier periods. Of course just the opposite occurs in periods of dry weather, when later rainfall is absorbed by the drainage area so that the discharge remains below normal.

Analysis shows that when the Hurst exponent H differs from 0.5, then the underlying process displays *persistence* on all time scales. In the case when H is greater than 0.5, if for some time in the past we had an increase in rainfall levels, then we will also see an increase in the future—on the average. Consequently, an increasing trend in the past implies an increasing trend in the future for all processes with H greater than 0.5. Moreover, this applies for arbitrarily large time horizons T. On the other hand, if the exponent H is less than 0.5, we have *antipersistence*. In these situations an increase in the past implies a *decrease* in the future.

Plotting the logarithm of the rescaled range R_T/S_T against the logarithm of the time interval T results in a straight line, the slope of which is precisely the Hurst exponent H. Figure 6.29 shows the results of such an exercise for water level minima on the river Nile over the years 622–1469. Here we see a value $H = 0.91$, strongly indicative of persistent flooding, which of course is exactly what was observed in these years.

Processes with $H \neq 0.5$, in which persistence or antipersistence appears, are called *fractional Brownian motions*. They have infinite

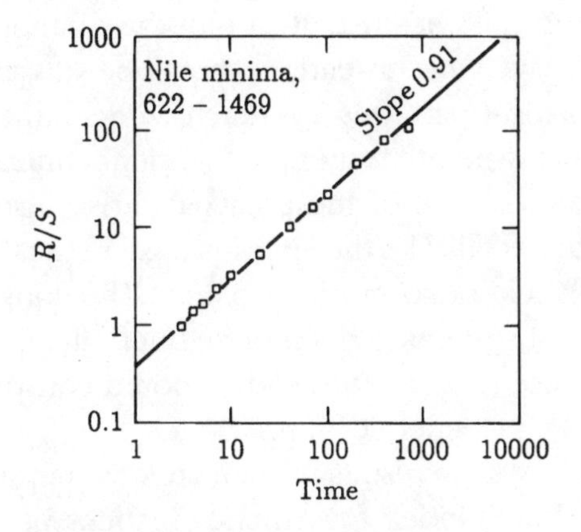

FIGURE 6.29.
HURST EXPONENT FOR WATER LEVELS ON THE NILE

long-run correlations. In particular, past increments are correlated with future increments. To those with a nose for money, this fact should set the alarm bells ringing since if price movements of speculative commodities also obey Hurst statistics, it might prove profitable indeed to know this. So we close this chapter with an examination of this very point.

Persistence in the Capital Markets

For analyzing price movements on stock and currency exchanges, it's technically convenient to use logarithmic returns rather than the actual prices. The logarithmic return at time t is defined as $s_t = \log(p_t/p_{t-1})$, where p_t is the actual price at time t. So the first step in performing a rescaled range analysis of a financial market is to convert the actual prices into logarithmic returns. Next we calculate the cumulative deviation and the range for several time increments T. For example, if we have a monthly time series involving 40 years of data, we might begin with $T = 6$-month increments. This would then divide the data into 80 independent, nonoverlapping periods. We then calculate the range and cumulative deviation from the average for each of these periods, obtaining 80 separate R/S observations. Averaging these 80 observations finally gives an R/S estimate for the series with $T = 6$ months.

Continuing this process for $T = 7, 8, \ldots 240$ months, we can study how the estimate of the Hurst exponent H changes with varying time horizons. Of course, we expect the estimate to fluctuate more as the time horizon increases since we have fewer observations to average.

To represent these results graphically, it's useful to plot the quantity log *T* versus log (*R/S*) since, as we've seen earlier, the slope of that curve is the number *H*. For the most part, this approach works rather well. But the reader should be warned that there are some technical subtleties that cannot be ignored. Some of these caveats arise from the fact that while mathematical fractals like the Sierpinski gasket scale forever, natural fractals like stock price curves do not. The long-memory process underlying most systems is not infinite but finite. For the more technical details of these matters, the reader should consult the references cited in the To Dig Deeper section.

The diagram in Figure 6.30 shows the results of such an *R/S* analysis for monthly returns on the S&P 500 index for American stocks over the 38-year period January 1950–July 1988. Here we see that the S&P 500 returns depart from the line $H = 0.78$ only after about 48 months ($\log_{10} 48 \approx 1.7$). This tells us that a long-memory process is at work for periods less than $T = 48$ months. After that point the graph begins to follow the line $H = 0.5$, corresponding to a purely random walk. Thus, returns that are more than 48 months apart have very little correlation left, on the average. And if you think this conclusion holds

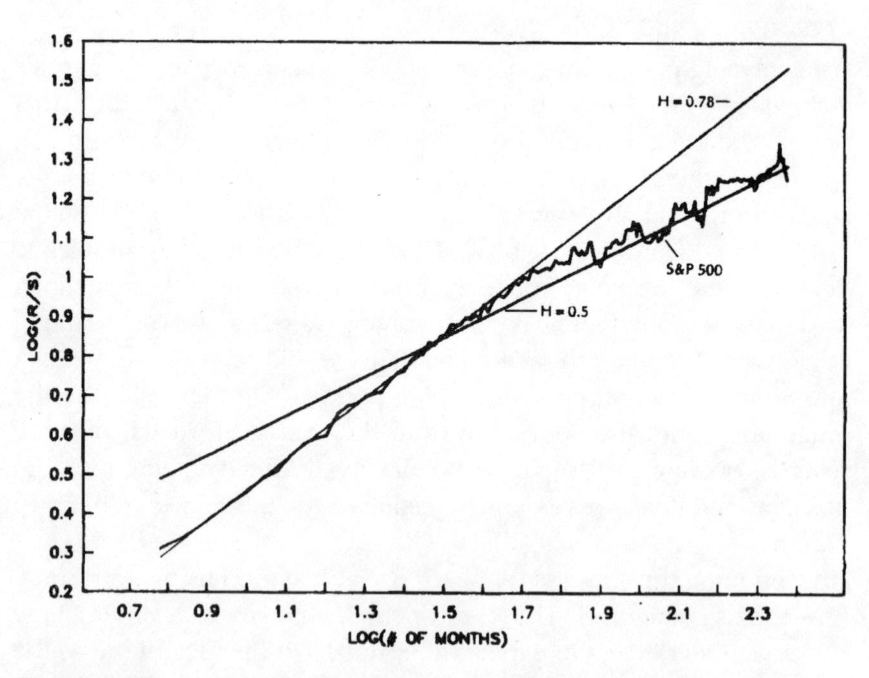

FIGURE 6.30. ESTIMATE OF THE HURST EXPONENT FOR MONTHLY S&P 500 RETURNS

just for an aggregate measure of the market like the S&P 500, have a look at Figure 6.31. This shows the same long-memory effect at work in the price history of IBM stock. Moreover, prices of other individual stocks ranging from high-tech issues like Apple Computer to ultraconservative utilities like Niagara Mohawk display the same fractal structure.

These high estimates for H provide strong support for the claim that the stock market is not a random walk, but rather is a fractal with trend-reinforcing behavior. This conclusion is in direct contradiction to the cherished Efficient Market Hypothesis (EMH), which describes the market as a roulette wheel with no memory (i.e., a wheel with mean-reverting behavior. The *R/S* analysis shows that the no-memory independence assumption underlying the EMH should be viewed with the kind of skepticism normally reserved for the no new taxes campaign promises of politicians.

In passing, let me note that fractal behavior is not confined to the American markets. Table 6.2 shows estimates of the Hurst exponent and cycle times for the German, Japanese and British stock markets, alongside the American. Each of them has an exponent far removed from the pure random walk level of $H = 0.5$. It's not just in stock markets where we see long-memory effects, either. The same phenomena occur in the international currency markets—usually. As an illustration, Figure 6.32 shows the *R/S* analysis for the daily exchange rate between the U.S. dollar and the Japanese yen during the period January 1973–December 1989. The same kind of picture arises with the dollar/mark and dollar/pound rates as well. However, the situation is rather different for the Singapore dollar.

The U.S. dollar/Singapore dollar rate is a true random variable, having a Hurst exponent $H = 0.5$. A little investigation into the situation

TABLE 6.2.
ESTIMATES OF *H* FOR INTERNATIONAL STOCK MARKETS

Market	*Hurst Exponent*	*Cycle Time (Months)*
S&P 500	0.78	48
Germany	0.72	60
Japan	0.68	48
Britain	0.68	30

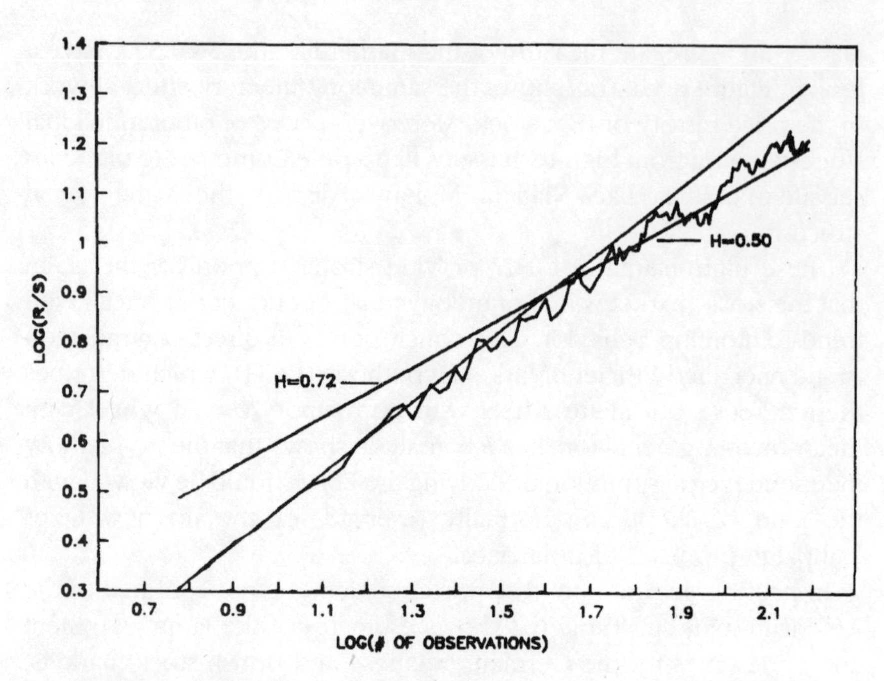

FIGURE 6.31. ESTIMATE OF THE HURST EXPONENT FOR MONTHLY IBM STOCK RETURNS

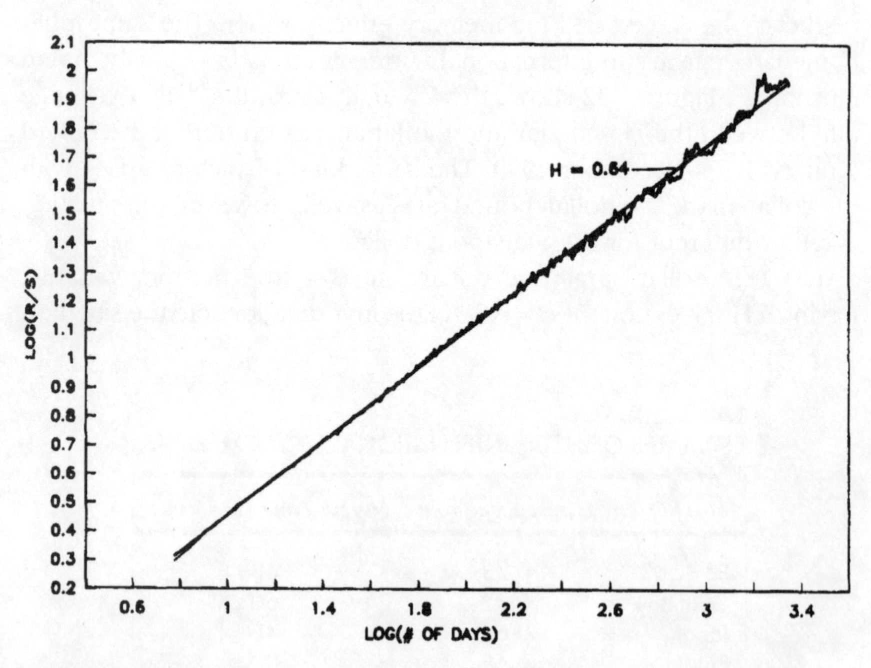

FIGURE 6.32. ESTIMATE OF THE HURST EXPONENT FOR DAILY YEN/DOLLAR EXCHANGE RATE

soon turns up the reason for this anomaly. The Singapore government deliberately manages their currency to track the U.S. dollar. As a result of this conscious effort, any fluctuation in the exchange rate is due to purely random fluctuations in the timing of the trades that fix the rate. So what at first glance looked like a violation of our "Fractal Market Hypothesis," turns out to be a validation of it instead!

There is much more than could (and should) be said about the fractal nature of speculative markets. Fortunately, much of it has already been said with great eloquence and detail in the papers and books cited in the To Dig Deeper section. So on this happy note we conclude our discussion of how emergent phenomena in both space and time lead to unexpected behavior of natural and human systems. The time has finally arrived to summarize all of our deliberations about why we should "expect the unexpected."

SEVEN

THE SIMPLY COMPLEX

On the Creation of a Science of Surprise

Reality is not perceived, it is conceived.

—C. S. Holling

Truth comes out of error more easily than out of confusion.

—Francis Bacon

Basic research is what I am doing when I don't know what I am doing.

—Werner von Braun

The Anatomy of Surprise

One-half of the 1972 Nobel Prize in Economic Science was awarded to Stanford economist Kenneth Arrow for his "pioneering contributions to general economic equilibrium theory and welfare theory." Left unmentioned in this citation was perhaps Arrow's most long-lasting contribution, the so-called Arrow Impossibility Theorem, which imposes severe constraints on the ways we can hope to divide up a society's resources in a "fair" manner. The problem addressed by Arrow's result involves a group of people who must collectively decide how to equitably parcel out the resources of the group. Arrow imposes a handful of reasonable conditions on the decision process, conditions such as if every person prefers choice A to choice B, then the group as a whole will prefer A to B and that there can be no single person whose choices dominate those of the group (no dictator). The Impossibility Theorem then states that there is no way for the group to arrive at a collective preference that won't violate one or another of the conditions. Put more succinctly, there can be no perfect form of government.

This kind of impossibility result illustrates the sort of surprise that arises out of subtleties of logic—subtleties that sometimes blow up in your face. The only way to get around these logical time bombs in this case is to drop one or more of the conditions, an act that would constitute a major step away from what many idealistic dreamers see as the very essence of the egalitarian and democratic ideal of total fairness. Arrow's result is not an isolated singularity, either. Here's another everyday setting where surprises of this kind can surface.

Each year professional football, basketball and baseball teams in the United States engage in a draft of new players, in which the teams choose players in reverse order to how the teams finished in the previous year's standings. Let's make the following assumptions about how teams behave in such a draft:

• *Strict preferences and partial ordering*—Each team has a strict preference ordering on the players considered individually. So, for instance, if a team has the preference ordering 123456 on the six players 1 through 6, then the subset of players {1,4,5} will be preferred to the subset {2,4,6}. But this ordering of the individual players generates a *partial ordering* by pairwise comparison on any subset of players the team might receive in the draft. This means that the team might not be able to rank order all possible subsets of players. For instance, the team will not be able to decide if it prefers the set of players {1,2,6} over the set {2,3,4}.

• *Self-interest*—Each team's goal is to benefit itself, not to hurt other teams. Consequently, no team will select a player solely to deny that player to another team.

- *Independence*—Each team makes its selections independently of the other teams. In other words, there are no coalitions or hidden agreements.
- *Complete information*—Each team knows the preference orderings of the other teams.

With the possible exception of the last assumption, each of these conditions seems to fit perfectly well with the circumstances underlying real sports drafts. For the sake of nomenclature, let's agree to call a team's choice of a player *sincere* if from the players not yet selected, it picks the highest-rated player on its preference list.

Assume initially that there are only two teams in the draft. Then there exists a systematic procedure (i.e., an algorithm) for generating sincere selections, which ensures that the resulting allocation of players *A* is optimal. This means that there is no other allocation of players *A′* in which each team either gets the same players as with *A* or prefers *A′* to *A* under pairwise comparison. In short, neither team would be better off under a different allocation. (Technically, this is what's termed a *Pareto optimal allocation*.) The sincere outcome is always optimal with regard to pairwise comparison—even when there are more than two teams. So if each team always chooses what it regards as the highest-ranking player left in the pool, the final outcome will be optimal in the foregoing Pareto sense. But what about real sports drafts, where there are usually many more than just two teams and in which the way the teams choose may not be sincere? That's where the fun and the surprises begin.

If there are three or more teams, use of optimal *sophisticated* play—choices that are not necessarily sincere—can lead to an allocation that's not optimal by pairwise comparison. Even more strongly, the outcome may be strictly *worse* than what each team would have received if they had all chosen sincerely. So if each team follows an individually optimal choice strategy, the final result can be what's called a *Prisoner's Dilemma situation*, in which *all* the teams are worse off than if they had each chosen sincerely. A corollary of this fact is that a team may do better for itself by occupying a later position in the draft. For an explicit example showing how this kind of dilemma can arise, the reader is referred to the To Dig Deeper section.

We have seen two simple choice situations—societal preferences and sports drafts—in which the logical structure of the decision process can lead to unexpected collective outcomes when each individual tries to stay true to his or her own preferences. Thus we conclude that a major source of surprise is simply our human inability to trace through the logical consequences of our assumptions. But byzantine chains of logical connection are just one of the surprise-generating mechanisms we've explored throughout the course of this book. Table 7.1 summarizes "all the usual suspects" in our search for the essence of surprise. Now let's revisit these surprise generators, both by way of summary and as a prelude to the consideration of what it would take to create a science of surprise.

Instability and Economies of Agglomeration

In the opening chapter, we saw how increasing returns to scale could lead to an inferior technology like the VHS videocassete format beating out a technically superior product like the Sony Betamax. But the kind of instability leading to this sort of market "tipping" is of far broader currency than merely as a way of settling technological wars of attrition. Here's an example of how the same phenomenon occurs in regional economics.

Ever since the ascendancy of high-tech areas like Silicon Valley in northern California and the Route 128 corridor outside Boston, towns throughout the industrialized world have grappled with the question of how to make the same lightning strike in their own backyard. Reduced to its essence, the fundamental puzzle these regional planners have to solve is When do economies of agglomeration lead to a "Silicon Valley"—a single dominant location or region that monopolizes an industry? And how do you go about creating circumstances that give rise to such an agglomeration?

TABLE 7.1.
SURPRISE-GENERATING MECHANISMS

Mechanism	***Surprise Effect***
Logical tangles	Paradoxical conclusions
Catastrophes	Discontinuity from smoothness
Chaos	Deterministic randomness
Uncomputability	Output transcends rules
Irreducibility	Behavior cannot be decomposed into parts
Emergence	Self-organized patterns

W. Brian Arthur has examined this question from the point of view of positive-feedback economics, coming to some extremely interesting and important conclusions. Basically, Arthur found that if the returns firms receive for locating in a particular area continue to increase without limit as more and more firms in the industry move in, then the industry will indeed always cluster in a single area. But the *particular* region the firms cluster in depends both on how attractive the region is and on the historical order of entry of the firms into the area. So there's a chance factor that enters into the selection process determining the specific town or region in which the entire industry settles.

On the other hand, if returns level off as more and more firms move into the area, then whether or not the industry will cluster in a particular region depends on the exact sequence in which the various firms enter the area. Moreover, certain sequences of entry can produce regional sharing of the industry as if there had been no agglomeration effects at all. What this adds up to is that increasing returns to scale do not guarantee a monopoly outcome unless the returns are unbounded.

This conclusion is a bit disheartening from a regional planning point of view since it says that unless a regional government is ready to offer massive and continuing financial incentives to firms agreeing to locate in their area, there is no guarantee of being able to transform the region into a new Silicon Valley. This example shows instability, the second of our surprise generators, coming to the fore. The instability in this case is the unstable nature of the attractors that arise from the locational process due to the system involving increasing returns to scale (i.e., positive-feedback loops).

Another major obstacle to the formation of faithful models of real-world phenomena lies in the fact that not all natural and human phenomena are the end result of following a set of rules. Surprises can happen when we deal with systems in which there is no computable way to mirror in symbols the unfolding of their dynamics. Let's look at an example illustrating this type of surprise.

The Wave Equation and Wave Motion

Whether it's the breaking of a wave on the beach or the transmission of a television signal through the atmosphere, there are few physical processes more familiar, or more important, than simple wave motion.

Mathematical physicists have made a tidy living for centuries by developing refined methods for solving the equations governing transmission, reflection and absorption of waves, be they vibrations of water, electromagnetic signals, bridge cables or a child's jump rope. Nevertheless, surprises are in store for those who believe that our mathematical models of wave behavior and the *real* motion of a *real* wave necessarily have anything to do with each other.

Recently, mathematicians Marian Pour-El and Jonathan Richards of the University of Minnesota have shown that under reasonable mathematical—but dubious physical—circumstances, solutions of the classical wave equation are uncomputable. To understand the meaning of this result, recall that the wave equation describes how a wave propagates through a region of space over the course of time. To get this wave motion started, we have to specify an initial waveform at time zero. For example, in order to get any sound to emerge from the string of a harp or guitar, we have to pull the string away from its rest position. The shape of the plucked string just before we let it go is the starting waveform, or initial state. From there, the wave equation will tell us what form the string assumes at any future time; hence, the tone that the string will produce.

Pour-El and Richards have shown that there exist perfectly good mathematical waveforms that lead to uncomputable solutions of the wave equation. So if our wavy system starts from one of these unlucky initial wave patterns, there is no computer program or algorithm that can track its future behavior. In short, the output of the mathematical model necessarily parts company from the actual behavior of the physical system—the very essence of what we mean by a surprise. Here we see a situation in which the underlying uncomputability of our model of the world creates an impassable barrier to the elimination of possible surprises. And this sort of uncomputability is not just confined to inanimate physics.

We noted in an earlier chapter arguments by Oxford physicist Roger Penrose against the possibility of there ever existing a genuine thinking machine. A key element in Penrose's anti-AI claim is that at least some human cognitive processes involve uncomputable operations. Since by definition a computer can generate only the values of computable functions, this says that the capabilities of the human brain must transcend that of any mere mechanism. So goes Penrose's argument, anyway. Of course, he is silent on the precise nature of these uncomputable operations, burying them in one of the least well-understood parts of modern physics: quantum fields. Such are the charms of unbridled speculation in modern science! So we find uncomputability as another major contender on our list of the root causes of surprise. Let's cross-examine our next suspect, irreducibility.

The Five-Body Problem

Certainly the most (in)famous problem of celestial mechanics involves the question of how bad the behavior of a collection of pointlike masses can be if each is influenced only by the gravitational attraction of the others. This is the well-known *N-Body Problem*. And the answer? Very bad indeed!

Recently, Zhihong Xia of the Georgia Institute of Technology has shown that a system of five or more point masses can behave so badly that within a *finite* time *all* the particles will fly off to infinity. Xia's result, of course, relates to a mathematical idealization of a system of real celestial bodies. For example, real planets are not geometrical points. Moreover, Newton's laws of gravitation are only an approximation to Einstein's general theory of relativity, which contains the equations governing actual planetary motion, especially for those objects moving near the Sun. Nevertheless, it's from exactly this kind of mathematical model that we form our expectations of what real particles or planets will be doing over the course of time.

From a reductionistic point of view, the natural temptation in studying the behavior of several bodies is to break the system up into subproblems involving a lesser number of bodies. And, in fact, it's possible to give a complete mathematical solution for what happens in the case of a two-particle system. So it's irresistibly tempting to try to solve, say, the three-body problem by piecing together somehow the solutions to three two-body problems. Almost from the time of Henri Poincaré and Paul Painlevé, who proposed the original N-Body Problem around the turn of the century, mathematicians have known that such reductionistic approaches cannot be made to work. The essence of the problem lies in the linkages (i.e., forces) among *all* the particles. As soon as you start ignoring any of these connections, you end up throwing out the problem with the bathwater, so to speak.

Xia's contribution was to exhibit a specific five-body system in which all five particles go off to infinity after a finite amount of time. This illustrates the perhaps depressing fact that if the degree of connectivity in the system is great enough, some pretty awful surprises can emerge from the behavior of a system having only a small number of particles. It's not necessary to have a large number of interacting particles for bad things to begin to happen; you just need the right (or, perhaps, wrong) kinds of interactions. And speaking of interactions, here comes our final contender in the surprise-generation game: emergence.

Genes and Kauffman Nets

Every human cell contains roughly 100,000 genes—including an unknown number of regulatory genes—all switching each other on and off in an unimaginably complicated network of interactions. Stuart Kauffman of the Santa Fe Institute is a theoretical biologist who's been working for the last twenty years trying to explain the puzzling fact that all this switching on and off doesn't lead to utter chaos, but rather results in the cell organizing itself into stable patterns of activity appropriate for its particular function in the organism. How is it that this seemingly random operation of individual genes leads the cell to configure itself into a stable, workable structure? Speaking in Darwinian terms, it seems difficult to see how new types of organisms could possibly arise out of merely random mutations and natural selection, which is the standard Darwinian party line. Something more seems to be needed to account for the great diversity of living forms surrounding us today. The answer, according to Kauffman, lies in the marked preference of complex systems to spontaneously organize themselves into persistent patterns of activity that work. As Kauffman puts it, "Darwin didn't know about self-organization."

In contrast to mainline biologists and chemists, who try to explain the emergence of new patterns by looking at genetic regulation in painstaking biochemical detail, Kauffman has built a mathematical model in the form of a network of interactions that mimic the genetic regulatory activity. Suppose we have a network of N genes, each regulated by K other genes. Thus, there are a total of 2^K possible inputs to each gene in the network. At each moment, Kauffman assumes that one of these input patterns is selected at random. Moreover, he also randomly chooses one of the 2^K possible ON-OFF patterns for each set of inputs to each gene. This pattern at each gene determines whether the gene will be ON or OFF at the next instant. In short, both the wiring diagram linking the genes and the logical rule of their operation are chosen at random. The reader will recognize these *Kauffman networks* as cellular automata of a slightly different type than those considered in the last chapter. For the genetic network, the rule of state transition is chosen randomly at each time moment, as is the "neighborhood" of each cell, which may be decidedly nonlocal for these types of nets.

In numerous experiments with different values of N and K, what Kauffman saw was not total chaos at all. Rather, such networks showed a powerful tendency toward self-organization by settling into a small number of different periodic attractors. For example, when $K = 2$, the length of these cycles, as well as the total number of different cycles, is small, typically on the order of $\sqrt{N}$ in both cases. Kauffman believes these cycles can be identified with the possible cell types that may arise from the genetic network. So a cell type is a stable recurrent pattern of gene expression created solely by the logical structure built into the genetic network.

In Kauffman nets, gene A turns on gene B, which then turns on gene C and inhibits gene D and so on. What this type of model shows is that stable cellular types may arise spontaneously as the attractors of a dynamical system. And it is through the subtle interplay between the stable and unstable attractors—cooperation and competition—that patterns of change and periods of stasis can slowly evolve. Furthermore, employing some of the arguments that we saw in the last chapter, Kauffman's models tell us that the formation of these stable patterns is almost inevitable, regardless of how disorganized the network is to begin with. Basically, the genetic interaction dynamics seem to force the cellular genome to spontaneously organize itself into a viable structure.

So there we have it: logical tangles and self-reference, chaotic motion, static instability, uncomputability, irreducibility, emergence—six quite different types of surprise generators, any one of which can and often does lead to models of reality departing in noticeably important ways from reality itself. Where this all seems to be leading is to the fact that both natural and human affairs are just plain complex. If we accept the provisional conclusion that it's the complex systems of nature and life that produce the surprises, we have little choice but to confront head-on the difference between the simple and the complex. But does it make any sense at all to speak of a "science of complexity."

"COMPLEXIFICATION"

A few years ago, I saw a cartoon showing two scientists arguing over the meaning of complexity. In suitably dogmatic terms, the first scientist asserted, "Complexity is what you don't understand." Responding to this temerarious claim, his colleague replied, "You don't understand complexity." This circular exchange mirrors perfectly to my eye how the informal term *complexity* has been bandied about in recent years—especially within the normally flinty-eyed community of system scientists—to characterize just about everything from anesthesiology to zymurgology. Here we want to explore just a few of the dimensions of the problem of trying "scientify" the simply complex.

Science-fiction writer Poul Anderson once remarked, "I have yet to see any problem, however complicated, which, when you looked at it the right way, did not become still more complicated." Substituting the word *complex* for *complicated*, this statement serves admirably to capture two of the key points needed to understand what's at issue in turning the casual, everyday notion of a complex system into something resembling a science.

First is the realization that complexity is an inherently subjective concept; what's complex depends upon how you look. When we speak of something being complex, what we're doing is making use of everyday language to express a feeling or impression that we dignify with the label *complex*. But the meaning of something depends not only on the language in which it is expressed (i.e., the code), the medium of transmission and the message, but also on the context. In short, meaning is bound up with the whole process of communication and doesn't reside in just one or another aspect of it. As a result, the complexity of a political structure, a national economy or an immune system cannot be regarded as simply a property of that system taken in isolation. Rather, whatever complexity such systems have is a joint property of the system *and* its interaction with another system, most often an observer and/or controller.

This point is easy to see in areas like economics and finance. For instance, an individual investor interacts with the stock exchange and thereby affects the price of a stock by deciding to buy or sell. This investor then sees the market as complex or simple, depending on how the prices are perceived to be changing. But the exchange itself acts upon the investor too, in the sense that what is happening on the floor of the exchange influences the investor's decisions. This *back interaction* causes the market to see the investor as having a certain degree of complexity, in that the investor's actions cause the market to be described in terms like *nervous, calm* or *unsettled*. This kind of two-way complexity becomes especially obvious in situations when the investor is one whose trades make noticeable blips on the ticker without actually dominating the market.

So just like truth, beauty, good and evil, complexity resides as much in the eye of the beholder as it does in the structure and behavior of a system itself. This is not to say that there do not exist *objective* ways to characterize some aspects of a system's complexity. After all, an amoeba is just plain simpler than an elephant by whatever notion of complexity you happen to believe in. The main point here is that these objective measures only arise as special cases of the two-way measures, in which the interaction between the system and the observer is much weaker in one direction than in the other.

The second key point brought out by Anderson's quotation is that common usage of the term *complex* is informal. The word is typically employed as a name for something that seems counterintuitive, unpredictable or just plain hard to pin down. So if it's a genuine *science* of complex systems we're after and not just anecdotal accounts based on vague personal opinions, we're going to have to translate some of these informal notions about the complex and the commonplace into a more formal, stylized language, one in which intuition and meaning can be more or less faithfully captured in symbols and syntax. The problem is that an integral part of transforming complexity (or anything else) into a science involves making that which is fuzzy precise, not the other way around, an exercise we might more compactly express as "formalizing the informal."

Just to bring home this point a bit more forcefully, let's pause for a moment to consider some of the properties associated with *simple* systems by way of inching our way to a feeling for what's involved with the complex. Generally speaking, simple systems exhibit the following characteristics:

• *Predictable behavior*—There are no surprises in simple systems; simple systems give rise to behaviors that are easy to deduce if we know the inputs (decisions) acting upon the system and the environment. If we drop a stone, it falls; if we stretch a spring and let it go, it oscillates in a fixed pattern; if we put money into a fixed-interest bank account, it grows to a predictable sum in accordance with an easily understood and computable rule. Such predictable and intuitively well-understood behavior is one of the principal characteristics of simple systems.

Complex processes, on the other hand, generate counterintuitive, seemingly acausal behavior that's full of surprises. Lower taxes and interest rates lead to higher unemployment; low-cost housing projects give rise to slums worse than those the "better" housing replaced; the construction of new freeways results in unprecedented traffic jams and increased commuting times. For many people, such unpredictable, seemingly capricious, behavior is the defining feature of a complex system.

• *Few interactions and feedback/feedforward loops*—Simple systems generally involve a small number of components, with self-interactions dominating the linkages among the variables. For example, primitive barter economies, in which only a small number of goods (food, tools, weapons, clothing) are traded, seem much simpler and easier to understand than the developed economies of industrialized nations, in which the pathway between raw material inputs and finished consumer goods follows a labyrinthine route involving large numbers of interactions between various intermediate products, labor and capital inputs.

In addition to having only a few variables, simple systems generally consist of very few feedback/feedforward loops. Loops of this sort enable the system to restructure, or at least modify, the interaction pattern among its variables, thereby opening up the possibility for a wider range of behaviors. To illustrate, consider a large organization that's characterized by variables like employment stability, substitution of capital for human labor, and level of individual action and responsibility (individuality). Increased substitution of work by capital decreases the individuality in the organization, which in turn may reduce employment stability. Such a feedback loop exacerbates any internal stresses initially present in the system, leading possibly to a collapse of the entire organization. This type of collapsing loop is especially dangerous for social structures, as it threatens their ability to absorb shocks, which seems to be a common feature of complex social phenomena.

- *Centralized decision-making*—In simple systems, power is generally concentrated in one or at most a few decision-makers. Political dictatorships, privately owned corporations and the Roman Catholic Church are good examples of this sort of system. These systems are simple because there is very little interaction, if any, between the lines of command. Moreover, the effect of the central authority's decision upon the system is usually rather easy to trace.

By way of contrast, complex systems exhibit a diffusion of real authority. Generally, such systems seem to have a nominal supreme decision-maker, but in actuality the power is spread over a decentralized structure. Actions of a number of units then combine to generate the actual system behavior. Typical examples of these kinds of systems include democratic governments, labor unions and universities. Such systems tend to be somewhat more resilient and stable than centralized structures because they are more forgiving of mistakes by any one decision-maker and are more able to absorb unexpected environmental fluctuations.

- *Decomposable*—Typically, a simple system involves weak interactions among its various components. So if we sever some of these connections, the system behaves more or less as before. Relocating American Indians to reservations produced no major effects on the dominant social structure in New Mexico and Arizona, for example, since, for various cultural reasons, the Indians were only weakly coupled to the dominant local social fabric in the first place. Thus the simple social interaction pattern present could be further decomposed and studied as two independent processes—the Indians and the settlers.

Complex processes, on the other hand, are irreducible. Neglecting any part of the process or severing any of the connections linking its parts usually destroys essential aspects of the system's behavior or structure. We have already looked at the N-Body Problem in this regard. Other examples include an electrical circuit, a Renoir painting or the tripartite division of the U.S. government into its executive, judicial and legislative subsystems. You just can't start slicing up systems of this type into subsystems without suffering an irretrievable loss of the very information that makes these systems a "system."

* * *

The foregoing points are pretty obvious, I think, and should hardly be matters of debate among the complex systems crowd. Nevertheless, it's from looking at the commonplace and the self-evident in new and interesting ways that new sciences emerge. Bridging the gap between the informal and the formal is a necessary first step in making something that passes for a science out of our intuitive, everyday feelings about the complex. But before entering into a discussion of just how this subjectivistic formalization might be carried out, let me pause for a moment to consider why we might want such a thing as a science of complexity in the first place.

As noted above, our impressions of complexity are something like an experience of meaning, part of a cultural cognitive map. And the meaning of our lives depends on the particular maps we use to decode our thoughts, choices and actions. But human societies have evolved to the point where the traditional maps no longer match our collective experience for very long. Thus, by coming up with a workable (i.e., scientific) theory of complexity, we can hope to be able to internally represent the experience of change by describing our collective reality as a process. This, in turn, would be a major step toward the development of a framework within which we can begin to understand how to control and manage what our maps tell us are complex processes.

A second, and somewhat more direct, reason for trying to create a science of the complex is to get a handle on the limits of reductionism as a universal problem-solving approach. When faced with a problem we don't understand, the traditional knee-jerk response is to invoke the old adage "When you don't know what to do, apply what you do know." Most of the time this translates into an attempt to decompose the hard problem into a collection of simpler subproblems that we understand. We then try to assemble the solutions of these bits and pieces into something that looks like an answer to the original question. Unfortunately, this procedure works just often enough to appeal to the prejudice of reductionists seeking rationalization for their particular brand of epistemological medicine.

But we're all familiar with examples like the behavior of gravitating bodies or, for that matter, the human body, in which any reductionistic approach of this sort irretrievably destroys the very nature of the problem. Such systems are complex. And it would be nice to have a theory tracing out the boundaries of the reductionistic approach, as opposed to blundering about like blind men, crashing up against these barriers before we even know they exist. So much for motivation. Now let's turn to the twin problems of formalization and objectification of the informal and subjective.

THE SCIENCE OF SURPRISE

The heart of the formalization process is shown schematically in Figure 7.1, which we might term the *modeling relation*. Here we see a natural (read: real-world) system *N* characterized by observations and relations stated in everyday language. The formalization process then involves encoding these characterizations of *N* into the symbols and strings of a formal logical (read: mathematical) system *F*. The key to understanding this process of formalization is to recognize that all notions of meaning (i.e., semantics) reside on the left-hand side of the diagram. So any real-world intuitions we have about *N*—its complexity, for example—belong to this side of the modeling relation. By way of contrast, there is no meaning at all on the right-hand side; *F* consists of mere abstract symbols, together with rules (a grammar) for how strings of these symbols can be combined to form new strings. Whatever meaning might inhere in these strings is then brought out by the decoding of the strings back into *N*. An example or two will fix this idea.

In Chapter Four we saw that during a course in mathematical logic at Cambridge University in 1935, Alan Turing was exposed to Hilbert's Decision Problem, which asks if there is any algorithmic procedure

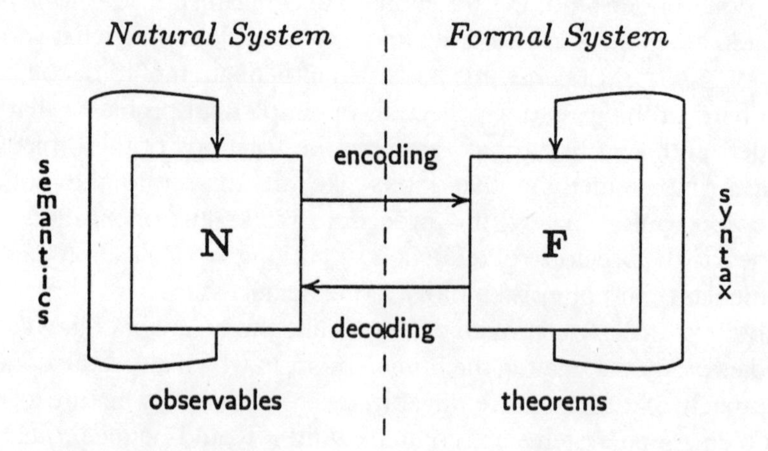

FIGURE 7.1. THE MODELING RELATION

for deciding whether a given statement made in a formal logical system is a theorem of that system. While trying to solve this problem, Turing invented the so-called Turing machine discussed earlier in our deliberations on computability. For our purposes here, what's important about this mathematical computer is that it represents the first successful attempt to formalize the informal notion of what it means to carry out a computation. Despite the fact that people had been calculating for several thousand years, it was not until Turing's work, less than sixty years ago, that the bridge was crossed from the informal system N = *computation* to its formalization as F = *Turing machine*.

While Turing's result showing that all computers are created equal is of enormous conceptual significance, I think it's important to point out that very few computer designers, if any, rely upon this fact as they go about their daily chores. So the formalization of the informal idea of a computation has had very little practical impact on problems of modern computer design and operation, despite the fact that the Turing machine serves as the conceptual foundation for a large part of what we now call computer science. In passing, let me mention another example of the same sort, namely, Gödel's formalization of the informal notion of *truth*. Again, not many mathematicians lose any sleep over the implications of Gödel's result for their work. Nevertheless, it's hard to deny the significance of incompleteness as we ponder the soundness of the mathematical enterprise. So it shouldn't come as much of a surprise were we to discover that a successful formalization of complexity will be equally useless from a practical point of view, yet equally profound from the standpoint of setting the foundations for a general theory of models.

Arguing by analogy from the historical genesis of these examples, we find that the formalization of the idea of complexity reduces to finding a symbolic structure in F to mirror our informal ideas about what it is that makes a system complex. In Turing's case, everyday ideas about what it means to carry out a computation were mirrored in the operations of the Turing machine, while Gödel mirrored what we think of as real-world truth by the idea of a mathematical proof. Almost all attempts made to carry out this kind of mirroring for complexity—information content, length of minimal computer programs, entropy, thermodynamic depth, to name but a few—come down to the translation of informally felt beliefs about the complex into formal, symbolic operations of one sort or another. But so far none of these formal surrogates has achieved a consensus in the system modeling community as being the "right" formalization. And despite the numerous interesting and technically deep results that have come out of these attempts, I can't think of a single system modeler whose work is influenced in the slightest by any of these characterizations (which, again arguing by analogy with Turing and Gödel, could be taken as an indicator that they are on the right track after all, I suppose). This sad fact indicates that there's something missing from these formalizations. Recalling the discussion given earlier, let me now argue that the missing ingredient is the explicit recognition that system complexity is a subjective, not an objective, property of an isolated system. But it can become objective, once our formalism takes into account the system with which our target system interacts.

Descriptions, Bifurcations and Complexity

Consider a system *N* and an observer who interacts with *N*. (Note: Here I'll use the emotionally laden term *observer* in the weakest possible sense to mean simply some other system that interacts in some way with the target system *N*, and not in the stronger sense of a system that measures or sees an attribute of *N*.) The observer creates a linguistic description of the system in the real world. This description is then formalized into a description in the mathematical world *F* by the process just discussed. We now ask, How many *inequivalent* descriptions of *N* can our observer generate? My claim is that the complexity of the system *N as seen by the observer* is directly proportional to the number of such descriptions. Here's why.

Suppose our system *N* is a stone on the street. To most of us, this is a pretty simple, almost primitive kind of system because we are capable of interacting with the stone in a very circumscribed number of ways. We can break it, throw it, kick it—and that's about it. Each of these modes of interaction represents a different (i.e., inequivalent) way to interact with the stone. But if we were geologists, then the number of different kinds of interactions available to us would greatly increase. In that case, we could perform various sorts of chemical analyses on the stone, use carbon-dating techniques on it, x-ray it and so on. For the geologist, the stone becomes a much more complex object as a result of these additional—and inequivalent—modes of interaction. We see from this example that the complexity of the stone is a relative matter, dependent on the nature of the system with which the stone is interacting. And this idea is perfectly general, applying not just to stones on the street but to all systems. So how do we get a handle on the number of such inequivalent descriptions available to a particular observer?

Recall that the observer begins by creating an informal description of a system *N* in the real world. The observer must then encode this description into the symbols and strings of a formal logical structure *F*. Deciding whether or not two informal, real-world linguistic descriptions are equivalent is a pretty tricky affair, opening up all sorts of depressing debates and semantic confusions of the kind that permeate the arts and humanities. But not so in the pristine world of formal logical systems. Every formal mathematical structure in *F* comes equipped with its own natural notion of equivalence, a notion that can be used to classify the informal descriptions.

The idea underlying virtually all of these equivalence concepts is that two objects are taken to be equivalent if they can be transformed one to the other by a simple relabeling of the variables used to describe them in *F*. In short, two objects are equivalent if they differ only in the way we look at them (i.e., by a change of coordinates). So once we have coded our informal description of *N* into some formal mathematical structure like a set of differential equations, a directed graph, a collection of simplicial complexes or whatever, the natural equivalence relations for that type of structure can be employed to characterize the level of complexity of the system *N*. In essence, the complexity level is directly related to the number of equivalence classes that the observer creates by means of the natural equivalence relations defined for the coded version of *N* in *F*.

The foregoing idea also provides us with a way to identify when the complexity level shifts as we move through the space of descriptions. Additional complexity appears whenever one description *bifurcates* from another. So it's the bifurcation points in *F* that we can identify with increased complexity and, even more generally, with emergent phenomena. This observation enables us to relate the complexity of a system to things as seemingly diverse as the elementary catastrophes or the bifurcation points of vector fields.

The foregoing arguments lead inexorably to the claim that for complexity to become a science it's necessary—but far from sufficient—to formalize our intuitive notions about complexity in symbols and syntax. Furthermore, it's necessary for any such formalization to respect the fact that complexity is a subjective concept. One way to do this is to focus attention on the fact that no system lives in isolation. There are always other systems like observers or controllers that are responsible for deciding upon the particular formalization to be used. And, in fact, it is actually these systems that ultimately render the verdict as to what is and isn't complex.

So we come to the perhaps not so surprising conclusion that the creation of a science of complex systems is really a subtask of the more general, and much more ambitious, program of creating a theory of models. Complexity—as a science—is merely one of the many rungs on this endless ladder. As a call to arms in this battle with the complex, I can hardly do better than to close our deliberations on the sources and science of surprise with the following admonition from Marguerite Yourcenar, who in 1980 became the first woman writer elected to membership in the Académie Française. In her novel *Memoirs of Hadrian*, Yourcenar writes:

> The rules of the game: learn everything, read everything, inquire into everything. . . . When two texts, or two assertions, or perhaps two ideas, are in contradiction, be ready to reconcile them rather than cancel one by the other; regard them as two different facets, or two successive stages, of the same reality, a reality convincingly human just because it is complex.

TO DIG DEEPER

CHAPTER ONE

In the Beginning Is the Wor(l)d

For a brief account of Teigen's experiment with proverbs, see the entertaining volume

Kohn, A. *You Know What They Say. . . .* New York: Harper Perennial, 1990.

For the details, the original research article should be consulted. It's found in

Teigen, K. "Old Truths or Fresh Insights?" *British Journal of Social Psychology* 25 (1986), 43–49.

Edward Tenner's insightful remarks about the "revenge effect" can be found in

Tenner, E. "The Real World Takes Revenge on Planners." *International Herald Tribune,* July 31, 1991, 5.

For an account of the sandpile puzzle, see

Watson, A. "The Perplexing Puzzle Posed by a Pile of Apples." *New Scientist,* December 14, 1991, 19.

This article also points out that Liffman's computer model only slices the real sandpile into a two-dimensional planar section, and that extending it to three dimensions may well provide the missing link to explain the central dip in pressure.

Wittgenstein's work has been recounted in many places. One of the best is the prize-winning biography

Monk, R. *The Duty of Genius*. London: Jonathan Cape, 1990.

The relationship between a real-world phenomenon and its mathematical representation is explored in painstaking detail in the two-volume text

Casti, J. *Reality Rules: Picturing the World in Mathematics. I—The Fundamentals, II—The Frontier*. New York: Wiley, 1992.

Rules of Reality

A layman's account of how science stands today vis-à-vis the problems of prediction and explanation of everyday events is given in

Casti, J. *Searching for Certainty: What Scientists Can Know About the Future*. New York: Morrow, 1991 (paperback edition: Quill, New York, 1992).

A detailed critique of the idea that economics is in any way scientific is given in the volume

Rosenberg, A. *Economics—Mathematical Politics or Science of Diminishing Returns?* Chicago: University of Chicago Press, 1992.

Patterns, Puzzles and Paradoxes

For a fascinating discussion of paradoxes of all sorts, including the Penrose impossible staircase, Escher engravings and much, much more, see

Faletta, N. *The Paradoxicon*. New York: Doubleday, 1983.

In this same connection, see the book

Hughes, P., and G. Brecht. *Vicious Circles and Infinity*. London: Penguin, 1975.

The history and resolution of the Alabama Paradox is treated in considerable detail in the volume

Brams, S. *Paradoxes in Politics*. New York: The Free Press, 1976.

The counterintuitive network of springs, strings and weights was first presented in

Cohen, J., and P. Horowitz. "Paradoxical Behavior of Mechanical and Electrical Networks." *Nature* 352 (August 22, 1991), 699–701.

For a follow-up describing the Kansas City hotel incident, see

Podell, H. "Real-life Failure." *Nature* 355 (February 20, 1992), 683.

The psychological experiments by Herrnstein and Mazur involving inconsistency and irrationality in economic decisions are described in the popular article

Herrnstein, R., and J. Mazur. "Making Up Our Minds." *The Sciences* 27, no. 6 (1987), 40–47.

A famous article in which many of the same issues are raised is

Tversky, A., and D. Kahneman. "The Framing of Decisions and the Psychology of Choice." *Science* 30 (January 1981), 453–458.

It's All in the Motion

For the more technically inclined, a reasonably accessible account of dynamical systems, vector fields, attractors and all the rest is available in Chapter Two of the first Casti volume cited above. For more advanced technical treatments, see

Irwin, M. *Smooth Dynamical Systems.* New York: Academic Press, 1980.

Guckenheimer, J. and P. Holmes. *Nonlinear Oscillations, Dynamical Systems, and Bifurcations of Vector Fields.* New York: Springer, 1983.

Lichtenberg, A. and M. Lieberman. *Regular and Stochastic Motion.* New York: Springer, 1983.

Jackson, E. Atlee. *Perspectives of Nonlinear Dynamics.* Vols. 1 & 2. Cambridge: Cambridge University Press, 1990.

Chaos has had such public visibility that it's now in danger of becoming positively fashionable. For an introductory account, see the chaos "bibles"

Gleick, J. *Chaos.* New York: Viking, 1987.

Stewart, I. *Does God Play Dice?* Oxford: Basil Blackwell, 1989.

The work by Cesaré Marchetti showing the ubiquity of the logistic law is discussed with many details in

Marchetti, C. "Stable Rules in Social Behavior." *IBM Conference,* Brazilian Academy of Sciences, Brasilia, 1986.

For a popular account of this circle of ideas, see the recent work

Modis, T. *Predictions.* New York: Simon & Schuster, 1992.

For a much more thorough account of how positive feedback pervades economic processes, see

Arthur, B. "Positive Feedbacks in the Economy." *Scientific American,* February 1990, 94–99.

Helpman, E. and P. Krugman. *Market Structure and Foreign Trade.* Cambridge, Mass.: MIT Press, 1985.

Arthur, B. "Competing Technologies, Increasing Returns, and Lock-In by Historical Events." *The Economic Journal* 99 (1989), 116–131.

CHAPTER TWO

Continuity and Common Sense

For recent ammunition supporting the growing belief that the Chicxulub impact crater is the "smoking gun" responsible for the demise of the dinosaurs, see the article

"Huge Impact Tied to Mass Extinction." *Science* 257 (August 14, 1992), 878–879.

An excellent popular exposition of the meteorite-impact theory of the death of the dinosaurs, along with an enormously insightful account of the ways of science, is found in the volume

Raup, D. *The Nemesis Affair.* New York: Norton, 1986.

Ascertaining the exact nature of the relationship between a real-world phenomenon and a mathematical model of that phenomenon is probably the most fundamental epistemological problem in theoretical science. In fact, this same question shows up in many guises elsewhere, as well. The relationship between a painting of, say, a bowlful of fruit and the actual object, or the relationship between the word for *water* and the wet stuff itself are two simple examples. Philosophers, linguists, art critics and semioticians, among others, have debated the nature of this symbolic representation for centuries. So it should be no surprise to find a division of opinions on the matter even within the confines of the narrow world of science. Here are four somewhat eclectic but representative samples from the literature on this point:

Barbour, I. *Myths, Models, and Paradigms.* New York: Harper & Row, 1974.

Campbell, N. *What Is Science?* London: Methuen, 1921 (Dover reprint edition, New York, 1953).

Hempel, C. *Philosophy of Natural Science.* Englewood Cliffs, N.J.: Prentice-Hall, 1966.

Salmon, W. *Scientific Explanation and the Causal Structure of the World.* Princeton: Princeton University Press, 1984.

Although not directed toward the theme of scientific models and their connection with reality, a volume not to be missed as an account of the general symbol-versus-symbolized issue is the following account by Michel Foucault of the work of the Belgian painter René Magritte:

Foucault, M. *This Is Not a Pipe.* Berkeley, Calif.: University of California Press, 1983.

On matters of prediction and explanation—scientific style—via the medium of a set of rules, an introductory account for the general reader is

Casti, J. *Searching for Certainty: What Scientists Can Know About the Future.* New York: Morrow, 1991 (paperback edition: Quill, New York, 1992).

The Fall of the Wall and the Collapse of a Beam

Zeeman's model for the discontinuous change of a political ideology was published in

Zeeman, E. C. "A Geometrical Model of Ideologies," in *Transformations: Mathematical Approaches to Cultural Change,* C. Renfrew and K. L. Cooke, eds. New York: Academic Press, 1979, 463–479.

For the technically inclined, the family of functions so central to the catastrophe theory framework often arises in a somewhat different fashion than the route described in the text. Suppose we begin with a *single* function $f(x_1, x_2, \ldots x_n)$ of n variables. If the Hessian matrix of this function has rank less than n at the origin, then the origin is a degenerate critical point of f. This means that the second derivative of f does not completely characterize the local behavior of the function near the critical point at the origin. In short, the function f is unstable in the space of smooth functions, in exactly the sense described in the text. For example, the function $f(x) = x^3$ can be perturbed to a function with none or two critical points near the origin by means of an arbitrarily small change (e.g., $x^3 \rightarrow x^3 + ax$, $a > 0$ gives no critical points, while $a < 0$ gives two). What we want to do is find the *simplest* family of functions that contains f and is stable as a family.

To find this simplest family, we have to make use of a lot of very high-powered techniques and arguments from differential topology, singularity theory and other branches of mathematics. When all the smoke clears away, what we're left with is a parameterized family of the sort discussed in the text. For a fairly complete account of how to find the stable family for a given f, see Chapter Two of

Casti, J. *Reality Rules: Picturing the World in Mathematics. I—The Fundamentals.* New York: Wiley, 1992.

Two other fairly accessible, though still mathematical accounts of the ideas and machinery of catastrophe theory are

Poston, T., and I. Stewart. *Catastrophe Theory and Its Applications.* London: Pitman, 1978.

Saunders, P. *An Introduction to Catastrophe Theory.* Cambridge: Cambridge University Press, 1980.

The beam-buckling example is treated in great detail in the two books above. The original account, however, is

Zeeman, E. C. "Euler Buckling," in *Structural Stability, the Theory of Catastrophes and Applications in the Sciences.* Lecture Notes in Mathematics, Vol. 525. New York: Springer, 1976, 373–395.

The Magnificent Seven

For an introduction to Whitney's famous result involving mappings of the plane to the plane, see the Casti book cited in the preceding section. For a detailed account of the mathematics of this result, and for much additional information about catastrophe theory, see

Lu, Y.-C. *Singularity Theory and an Introduction to Catastrophe Theory.* New York: Springer, 1976.

A complete proof of the Classification Theorem, together with a wealth of good, bad and infamous applications of the theory, is available in the volume

Zeeman, E. C. *Catastrophe Theory: Selected Papers, 1972–1977.* Reading, Mass.: Addison-Wesley, 1977.

Physics and Metaphysics

Thom's interest in the work of Waddington leading up to the mathematics of catastrophe theory is recounted by Thom himself in the volume

Thom, R. *Mathematical Models of Morphogenesis.* Chichester, England: Ellis Horwood, 1983.

This volume is also a primary source for Thom's "metaphysical" way of applying catastrophe theory in biology and linguistics. The source that got everything started, however, was Thom's famous book on the topic, the French original of which appeared in 1972. The precise reference for the English translation is

Thom, R. *Structural Stability and Morphogenesis.* Reading, Mass.: Addison-Wesley, 1975.

An earlier paper by Thom that set the stage for much of the later biological speculations is

Thom, R. "Topological Models in Biology." *Topology* 8 (1969), 313–335.

Berry and Nye's striking applications of catastrophe theory to the problem of triple junctions for caustics are reported in

Berry, M., and J. Nye. "Fine Structure in Caustic Junctions." *Nature* 267 (1977), 34–36.

The famous work by D'Arcy Thompson in which he almost single-handedly created the field of relational biology is

Thompson, D'Arcy. *On Growth and Form.* Cambridge: Cambridge University Press, 1917.

For an up-to-date account of morphogenesis for the general reader, including a discussion of Turing's reaction-diffusion model for pattern formation, see Chapter Three of

Casti, J. *Searching for Certainty: What Scientists Can Know About the Future.* New York: Morrow, 1991 (paperback edition: Quill, New York, 1992).

The model of collapse of ancient civilizations is adapted from the original, which was presented in

Renfrew, C. "Systems Collapse as Social Transformation: Catastrophe and Anastrophe in Early State Societies," in *Transformations: Mathematical Approaches to Cultural Change,* C. Renfrew and K. L. Cooke, eds. New York: Academic Press, 1979, 481–506.

For an enlightening critique of Renfrew's model, along with an alternate approach to the problem of system collapse based on more traditional dynamical systems lines, see the fascinating work

Lowe, J. *The Dynamics of Collapse: A Systems Simulation of the Classic Maya Collapse.* Albuquerque: University of New Mexico Press, 1985.

A volume that takes up the entire question of social collapse in ancient societies from several complementary points of view is

The Collapse of Ancient States and Societies, N. Yoffee and G. Cowgill, eds. Tucson: University of Arizona Press, 1988.

For a more extended discussion of the physical versus metaphysical ways of catastrophe theory, the reader is invited to consult Chapter Two of

Casti, J. *Reality Rules: Picturing the World in Mathematics. I—The Fundamentals.* New York: Wiley, 1992.

The Theater of the Absurd

The main difference between controversies in science and those in the arts is that the scientific variety revolve by and large about conflicting interpretations of observations and theories rather than focusing on personalities and professional positions. A good account of some scientific controversies underscoring this point is given in the volume

Scientific Controversies. H. T. Englehardt, Jr. and A. Caplan, eds. Cambridge: Cambridge University Press, 1987.

For a very illuminating discussion of catastrophe theory in general, including probably the best account available of the controversy for the layman, see

Woodcock, A., and M. Davis. *Catastrophe Theory.* New York: E. P. Dutton, 1978.

The article by Gina Kolata that really brought the controversy to a head was published as

Kolata, G. "Catastrophe Theory: The Emperor Has No Clothes." *Science* (April 15, 1977). See also the correspondence in the issues of June 17 and August 26, 1977.

The "devastating" critique by Sussman and Zahler was published as

Zahler, R. and H. Sussman. "Claims and Accomplishments of Applied Catastrophe Theory." *Nature,* October 27, 1977. See also correspondence in the issues of December 1 and December 29, 1977.

CHAPTER THREE

Expecting the Unexpected

The issue of determinism versus predictability has been a staple of philosophers of science for generations. Discussions from various points of view can be found in the book of readings

Philosophy of Science. E. Klemke, R. Hollinger and A. D. Kline, eds. Rev. ed. Buffalo, N.Y.: Prometheus Books, 1988.

Of particular interest in the above volume is the article by Paul Thagard on why astrology is a pseudoscience. Other works worth consulting include

Kemeny, J. *A Philosopher Looks at Science.* Princeton: Van Nostrand, 1959.

Suppes, P. *Probabilistic Metaphysics.* Oxford: Basil Blackwell, 1984.

Recipes for Randomness

For a more detailed account of the equivalence between a computer program, a set of logical rules (a formal deductive system) and a dynamical system, see Chapter Nine of

Casti, J. *Reality Rules: Picturing the World in Mathematics. II—The Frontier.* New York: Wiley, 1992.

The problem of stability is probably the most studied aspect of dynamical systems—and with good reason. In fact, it's hard to imagine how any investigation of dynamics can get started before the stability properties of the system are understood. Of particular interest nowadays is the problem of structural stability. The basic setup is that we have a *family* of vector fields, often specified by different values of a parameter in a dynamical system. What we'd like to know is whether or not there is a discontinuous shift in the nature of the attractors as we move through family. In general, the answer is yes. So the structural stability question is For which members of the family do qualitative changes in long-run behavior occur? These family members are technically termed *bifurcation points.* For a more detailed consideration of these matters, see the texts

Casti, J. *Reality Rules: Picturing the World in Mathematics. I—The Fundamentals.* New York: Wiley, 1992.

Guckenheimer, J. and P. Holmes. *Nonlinear Oscillations, Dynamical Systems, and Bifurcations of Vector Fields.* New York: Springer, 1983.

The butterfly effect is described in many popular accounts of chaos. Two of the most easily accessible are

Gleick, J. *Chaos.* New York: Viking, 1987.

Stewart, I. *Does God Play Dice?* Oxford: Basil Blackwell, 1989.

The following volume is notable for its exploration of chaotic phenomena from a syntactic and semantic, as well as pragmatic, point of view. While somewhat more technical than the Gleick and Stewart books, its wide range of themes and novel applications in linguistics, genetics, cognition and game theory make it well worth the extra effort.

Nicolis, J. *Chaos and Information Processing.* Singapore: World Scientific, 1991.

The logistic rule has been studied to death in a myriad of different contexts. One of the most useful in bringing out the random nature of a system's behavior turns out to be casting the equation in the form of a coin-tossing experiment. By means of a bit of mathematical trickery, it can be shown that the logistic rule of the text is completely equivalent to the following "tent map":

$$x_{t+1} = \begin{cases} 2x_t & \text{for } 0 \leq x_t \leq \frac{1}{2} \\ 2(1 - x_t) & \text{for } \frac{1}{2} < x_t \leq 1 \end{cases}$$

Starting this process with any number x_0 between 0 and 1, we label the successive iterates with an H for Heads if it is between 0 and ½ and with a T for Tails if it's greater than or equal to ½. In this way, the behavior of the sequence can be thought of as a sequence of tosses of a coin.

Statistical analysis of a Heads/Tails sequence generated in this way turns up the by now not-so-surprising fact that the tools and tricks of the classical statistician offer no way to tell us whether the sequence came from the toss of a fair coin or, alternately, arose as the output of the completely deterministic process above. For further discussion of these matters, see the volumes

Lichtenberg, A. and M. Lieberman. *Regular and Stochastic Motion.* New York: Springer, 1983.

Jackson, E. Atlee. *Perspectives of Nonlinear Dynamics. Vol. 2.* Cambridge: Cambridge University Press, 1991.

For further details on the logistic map, including its interpretation in both population and atmospheric dynamics, see

The New Scientist *Guide to Chaos.* N. Hall, ed. London: Penguin Books, 1992.

Jensen, R. "Classical Chaos." *American Scientist* 75 (March–April 1987), 168–181.

Statistically Speaking

For the more technically inclined, a reasonably accessible account of the various tests for chaos is given in Chapter Four of the Casti volume *Reality Rules—II* cited for the preceding section. For other treatments, including original articles on the Period-Three Theorem, correlation dimensions and the like, see the collections

Chaos. Hao Bai-Lin, ed. Singapore: World Scientific, 1984.

Universality in Chaos. P. Cvitanović, ed. Bristol, England: Adam Hilger, 1984.

The most characteristic feature of chaotic systems is that points starting out close together diverge over the course of time. From an information-theoretic point of view, this divergence creates information: points that were initially indistinguishable become distinguishable as time unfolds. This simple observation leads to a test for chaos involving what's termed a *K-flow.*

Suppose now we think of the initial state of the system as being a bundle of nearby points, rather than a single point. The figure below shows that for a simple system, (a), the bundle retains its shape and moves in a regular manner over a restricted region of the state space. On the other hand, for the *ergodic* system, (b), the bundle again retains its general shape but now moves all over the state space. Finally, in the chaotic situation, (c), we see the bundle spread out like a drop of ink in water, eventually invading every part of the space. This spreading out occurs because trajectories in the bundle diverge from each other exponentially fast. Even more random than these *mixing ergodic* systems are the so-called *K-flows,* named after the famous Russian mathematician Andrei N. Kolmogorov. Their behavior is at the limit of unpredictability, in that having even an infinite number of measurements of where the system was in the past is of no help in predicting where it will be found next.

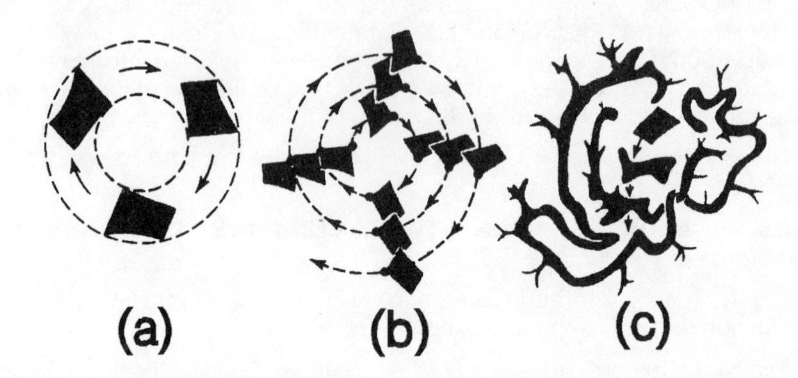

K-flows are widespread among systems in which collisions between particles dominate the dynamics. Simple examples of such processes include the case of three or more billiard balls in a box and the so-called classical gas, in which we regard each gas molecule as a hard little sphere. Although it has yet to be proved mathematically, the K-flow property is thought to characterize the vast majority of natural processes.

In order to test a set of data for the K-flow property, consider two trajectories starting from points extremely close together. Now introduce a quantity K, the so-called *K-entropy,* to measure the average rate at which the trajectories are moving apart from each other. Thus, K is the rate at which the trajectories are becoming distinguishable. It's evident that if the trajectories do not diverge at all, then $K = 0$, while chaotic dynamics leads to a positive, but finite, value for K. A totally random path, on the other hand, has $K = +\infty$.

Bulls, Bears and Beer

A layman's account of prediction and explanation as it pertains to the fluctuation of prices on stock exchanges is given in Chapter Four of

Casti, J. *Searching for Certainty.* New York: Morrow, 1991 (paperback edition: Quill, New York, 1992).

For more detailed accounts of how chaos intrudes into the pricing mechanism, see the recent accounts in

Peters, E. *Chaos and Order in the Capital Markets.* New York: Wiley, 1991.

Brock, W., B. LeBaron and D. Hsieh. *Nonlinear Dynamics, Chaos and Instability.* Cambridge, Mass.: MIT Press, 1991.

The workings of the Beer Game are described with great enthusiasm and detail in

Mosekilde, E., E. Larsen and J. Sterman. "Coping with Complexity: Deterministic Chaos in Human Decisionmaking Behavior," in *Beyond Belief: Randomness, Prediction and Explanation in Science,* J. Casti and A. Karlqvist, eds. Boca Raton, Fla.: CRC Press, 1991, 199–229.

Computing the Cosmos

Along with chaos came the breakdown of the perfect Laplacian correlation between predictability and determinism—even in principle. We've seen that a key feature of chaotic processes is the way that errors in assessing the initial state of a system grow exponentially over the course of time. Metaphorically speaking, the famed Argentine poet and writer Jorgé Luis Borges put it best in his short story "The Library of Babel." In that haunting tale there are an infinite number of hexagonally shaped cells in Borges's Library, these cells being arranged vertically one atop another in an infinitely long chain. This phantasmagorical Library contains every possible book that could be written, as well as leagues of strange volumes filled with random jumbles of letters and meaningless phrases. Chaotic systems correspond to those librarians of Babel who read every word and character in the books under their care. In contrast, nonchaotic systems are like readers who merely look at the titles and skim the contents. The sensitivity of chaotic systems to their initial conditions and, hence, their unpredictability comes from the fact that slight changes in the initial state correspond to different books in Borges's Library, which tell different tales. "The Library of Babel" can be found in the collection

Borges, J. *Labyrinths.* New York: New Directions, 1964.

In a footnote to this story, Borges speculates on a different kind of book having infinitely thin pages, densely packed like the rational numbers. With this book, each time you think you've turned a page it turns out that on closer inspection the page is itself a collection of thinner pages. This idea raises the question of a book with a *continuum* of pages. This theme is explored in the science-fiction novel *White Light* by author, logician and computer scientist Rudy Rucker.

CHAPTER FOUR

The Power of Paradox

A first-rate account of paradoxes of all sorts—logical, visual and otherwise—is given in the little volume

Faletta, N. *The Paradoxicon.* New York: Doubleday, 1983.

Another excellent introduction to these matters is

Poundstone, W. *Labyrinths of Reason.* New York: Doubleday, 1988.

A much more detailed discussion of legal paradoxes can be found in

Suber, P. *The Paradox of Self-Amendment.* New York: Peter Lang, 1990.

In this same regard, see Chapter Four in

Hofstadter, D. *Metamagical Themas.* New York: Basic Books, 1985,

which deals with Nomic, a game whose major activity focuses on how to change its rules lawfully.

A good introduction to Julesz figures is found in the volume

The Virtual Reality Playhouse. Corte Madera, Calif.: The Waite Group Press, 1992.

If you want to create your own Julesz figures, the N.E. Thing Corporation offers a low-cost computer package for IBM PCs that allows you to transform just about any planar figure into what they call a "Stare-EO." For more details, contact N.E. Thing Enterprises, Box 1827, Cambridge, MA 02139. It should be noted that psychologists have discovered that a small fraction of people (less than 10 percent) are neurologically wired in such a way that they cannot see the three-dimensional structure present in these figures. Too bad! So if you didn't "get" the message of Figure 4.3 despite hours of staring, it's possible that you may fall into this unlucky group.

Reality Rules

The logical conundrums involving George, Martha, Waldo and Myrtle were first presented in the vastly entertaining and educational work

Paulos, John A. *I Think, Therefore I Laugh.* New York: Columbia University Press, 1985.

An introductory account of formal systems can be found in

Levine, H. and H. Rheingold. *The Cognitive Connection.* New York: Prentice-Hall, 1987.

For a more technical account, emphasizing the connections between formal systems and languages, see the book

Moll, R., M. Arbib, and A. Kfoury. *An Introduction to Formal Language Theory.* New York: Springer, 1988.

For a vast amount of information about formal systems, decision procedures, Gödel's Theorems and much, much more besides, see the award-winning tome

Hofstadter, D. *Gödel, Escher, Bach: An Eternal Golden Braid.* New York: Basic Books, 1979.

For a proof of the decision procedure given in the text for the ★-✠-✩-system, see the article

Swanson, L. and R. McEliece. "A Simple Decision Procedure for Hofstadter's *MIU*-System." *Mathematical Intelligencer* 10, no. 2 (1988), 48–49.

Magic Machines and Busy Beavers

An excellent introductory account of the circle of problems surrounding computation, formal systems, Turing machines and the Halting Problem, as well as much more, is available in the article

Davis, M. "What Is a Computation?" in *Mathematics Today: Twelve Informal Essays,* L. A. Steen, ed. New York: Springer, 1978, 241–267.

For a general-readership development of the idea and workings of a Turing machine, see

Hoffman, P. *Archimedes' Revenge.* New York: Norton, 1988.

Rucker, R. *Mind Tools.* Boston: Houghton-Mifflin, 1987.

A more technical account of Turing machines and their connections with not only decision problems but also languages is available in the text

Davis, M. and E. Weyuker. *Computability, Complexity, and Languages.* Orlando: Academic Press, 1983.

A very stimulating collection of essays reviewing current knowledge about Turing machines and their many implications and ramifications in other areas is presented in

The Universal Turing Machine. R. Herken, ed. Oxford: Oxford University Press, 1988.

The problem of computability lies at the heart of the theory of computation. Excellent accounts of these matters are available in the volumes

Harel, D. *Algorithmics.* Reading, Mass.: Addison-Wesley, 1987.

Davis, M. *Computability and Unsolvability.* New York: McGraw-Hill, 1958 (expanded reprint edition: Dover, New York, 1982).

Boolos, G. and R. Jeffrey. *Computability and Logic.* 3rd ed. Cambridge: Cambridge University Press, 1989.

Epstein, R. and W. Carnielli. *Computability: Computable Functions, Logic and the Foundations of Mathematics.* Pacific Grove, Calif.: Wadsworth & Brooks/Cole, 1989.

Turing's original article, in which he introduced the idea of a Turing machine as a vehicle by which to discuss computability of numbers, is

Turing, A. "On Computable Numbers with an Application to the Entscheidungsproblem." *Proceedings of the London Mathematical Society,* ser. 2, vol. 42 (1936–37), 230–265; correction, vol. 43 (1937), 544–546.

The Busy Beaver Game was dreamed up by Tibor Rado of Ohio State University in the early 1960s. Compact introductory discussions of what's currently known about this problem and about properties of the Busy Beaver function *BB(n)* can be found in the article "Busy Beavers" in

Dewdney, A. *The Armchair Universe.* New York: Freeman, 1988.

Brady, A. "The Busy Beaver Game and the Meaning of Life," in *The Universal Turing Machine,* R. Herken, ed. Oxford: Oxford University Press, 1988, 259–277.

In 1973 Bruno Weimann discovered that the 4-state Busy Beaver can write thirteen 1s on the tape before halting. Thus, $BB(4) = 13$. So far no one knows the value $BB(5)$, although in 1984 George Uhing showed that $BB(5) \geq 1{,}915$. The program establishing this remarkable result is

	Symbol Read	
State	**0**	**1**
A	1, R, B	1, L, C
B	0, L, A	0, L, D
C	1, L, A	1, L, STOP
D	1, L, B	1, R, E
E	0, R, D	0, R, B

A relatively technical update of the current state of play of the Busy Beaver is given in the article

Machlin, R. and Q. Stout. "The Complex Behavior of Simple Machines," in *Emergent Computation*. S. Forrest, ed. Cambridge, Mass.: MIT Press, 1991, 85–98.

The Turing Machine Game is discussed in somewhat more detail in the survey article

Jones, J. "Recursive Undecidability: An Exposition." *American Mathematical Monthly,* September 1974, 724–738.

Here is the promised proof of the uncomputability of the winning strategy for Player B. Let $S(n)$ be the maximum number of steps that an n-state Turing machine can perform before halting. We must have $BB(n)$ less than or equal to $S(n)$ since the Busy Beaver function measures the number of 1s that can be printed, and it may be the case that on some steps the machine does not print a 1. By a strategy for Player B we mean a formula $m = f(n)$, specifying what number to select as a function of the number n chosen by Player A. By the rules of the Turing Machine Game, f is a winning strategy for Player B if and only if $S(n)$ is no greater than $f(n)$. Consequently, a winning strategy for Player B is just to take $f(n)$ to be the function $S(n)$ itself. However, it's eventually the case that $f(n)$ is less than the Busy Beaver function $BB(n)$, which in turn is less than or equal to $S(n)$. And this holds for every computable function f. Consequently, Player B has no *computable* winning strategy.

Ian Stewart has constructed the following slick proof showing the unsolvability of the Halting Problem. Suppose such a Halting Algorithm exists, and let d be the input data. Consider the following UTM program:

1. Check to see if d is the code for a UTM program P. If not, go back to the start and repeat.
2. If d is the code for a program P, double the input string to get $d \cdot d$.

3. Use the assumed Halting Algorithm for the UTM with input data $d \cdot d$. If it stops, go back to the beginning of this step and repeat.
4. Otherwise, halt.

Call the above program H. Now since H is a program, it has its own code h. Thus we can ask "Does H halt for input h?" It surely gets past step 1, since by definition h is the code for the program H. And H gets past step 3, as well, if and only if the UTM doesn't halt with input $h \cdot h$. Thus we conclude that H halts with input data h if and only if the UTM does not halt with input data $h \cdot h$. But the UTM simulates a program P by starting with the input data $P \cdot d$ and then behaving just like P operating on input data d. Therefore we see that P halts with input data d if and only if the UTM halts with input data $P \cdot d$. So if we put $P = H$ and $d = h$, then we find that H halts with input data h if and only if the UTM halts with input data $h \cdot h$—a direct contradiction to the result obtained a moment ago. Thus we conclude that there is no such Halting Algorithm.

A much more technical and philosophically oriented approach to the implications of the Turing-Church Thesis for both the psychology and the philosophy of mathematics is presented in the volume

Webb, J. *Mechanism, Mentalism, and Metamathematics.* Dordrecht, Netherlands: Reidel, 1980.

In this connection, see also

Arbib, M. *Brains, Machines, and Mathematics.* 2nd ed. New York: Springer, 1987.

The Turing-Church Thesis lies at the heart of the currently fashionable artificial intelligence debate, which revolves about the question of whether or not a computer can think like a human being. *If* human thought processes can be shown to all be "effective," and *if* the Turing-Church Thesis is correct, then it necessarily follows that there is no barrier, in principle, between the "thought processes" of machines and those of humans. But both of these ifs are very big ifs indeed, and no one has yet been able to give a knockdown argument resolving either half of this conundrum. For an account of the current state of play, as well as an extensive bibliography on the whole issue, see Chapter Five of

Casti, J. *Paradigms Lost: Images of Man in the Mirror of Science.* New York: Morrow, 1989 (paperback edition: Avon Books, New York, 1990).

For the inquisitive layman looking for a more detailed account of the players in the AI game, the following volume can be highly recommended:

Johnson, G. *Machinery of the Mind.* New York: Times Books, 1986.

Truth Is Stranger Than Proof

Gödel's results are recounted in many places, including the Hofstadter volume noted above. Other popular accounts include Chapter Six of

Casti, J. *Searching for Certainty: What Scientists Can Know About the Future.* New York: Morrow, 1991 (paperback edition: Quill, New York, 1992).

The following works also contain much material of general interest in this regard:

Rucker, R. *Infinity and the Mind*. Boston: Birkhäuser, 1982.

Hofstadter, D. "Analogies and Metaphors to Explain Gödel's Theorem." *College Mathematics Journal* 13 (March 1982), 98–114.

Nagel, E. and J. R. Newman. *Gödel's Proof*. New York: New York University Press, 1958.

An English translation of Gödel's pioneering paper, with an enlightening account of his life, can be found in the first volume of Gödel's collected works:

Kurt Gödel: Collected Works. Volume 1, S. Feferman et al., eds. New York: Oxford University Press, 1986.

Another source for the original paper, together with commentary, is

Gödel, K. *On Formally Undecidable Propositions of* Principia Mathematica *and Related Systems*. New York: Basic Books, 1962 (Dover reprint edition, 1992).

An assessment of Gödel's Theorem from a philosophical and mathematical point of view is contained in the collection of reprints

Gödel's Theorem in Focus. S. Shanker, ed. London: Croom & Helm, 1988.

People often wonder whether or not long-standing, seemingly intractable mathematical questions like Goldbach's Conjecture (every even number is the sum of two primes) are undecidable in the same way that Cantor's Continuum Hypothesis turned out to be undecidable. Musings of this sort give rise to the consideration of whether or not Gödel's results really matter to mathematics, in the sense that there are important mathematical questions that are truly undecidable. With the recent work of Chaitin and others, the comforting belief that there are no such problems seems a lot less comforting than it used to. For a discussion of some other "real" mathematical queries that are genuinely undecidable, see

Kolata, G. "Does Gödel's Theorem Matter to Mathematics?" *Science* 218 (November 19, 1982), 779–780.

Many details of Gödel's personality, views on life and philosophy, with an assessment of his mathematical and philosophical work, are found in the following book written by the well-known mathematical logician Hao Wang, who was a long-time acquaintance of Gödel:

Wang, H. *Reflections on Kurt Gödel*. Cambridge, Mass.: MIT Press, 1987.

Additional information about Gödel's life is given in

Dawson, J. "Kurt Gödel in Sharper Focus." *Mathematical Intelligencer* 6, no. 4 (1984), 9–17.

Kreisel, G. "Kurt Gödel: 1906–1978." *Biographical Memoirs of Fellows of the Royal Society* 26 (1980), 148–224.

The ideas underlying Gödel numbering are given in the Nagel and Newman book noted above.

Out-Gödeling Gödel

The complete story of Chaitin's independent discovery of algorithmic complexity and its connection with randomness is contained in his collection of papers

Chaitin, G. *Information, Randomness, and Incompleteness*. 2nd ed. Singapore: World Scientific, 1990.

Quite independently, the famous Russian mathematician Andrei Kolmogorov also hit upon the idea of defining the randomness of a number by the length of the shortest computer program required to calculate it. His ideas were presented in

Kolmogorov, A. "Three Approaches to the Quantitative Definition of Information." *Problems in Information Transmission* 1 (1965), 3–11.

The original formulation of Berry's Paradox involved a statement like "The smallest number that cannot be expressed in fewer than thirteen words." Since the preceding phrase contains twelve words, the paradox follows for exactly the same reasons as given for the more general phrase used in the text. A fairly complete account of the Berry Paradox and its relationship to complexity and Gödelian logic is available in the Rucker book *Infinity and the Mind* already noted. This volume also contains the background assumptions supporting Rucker's claim that 3 billion is an upper limit to the complexity of phenomena that the human mind can rationally encompass and comprehend.

As promised in the text, here is an outline of the proof showing that for sufficiently large N, there can be no proof that a particular string has complexity greater than N. Suppose we have a binary string that we suspect of having complexity greater than some fixed level N and we want to prove it. Assume such proofs exist. Then, since it takes only $\log N$ symbols to represent a number having magnitude N, we can use a program of length $\log N + K$ to search for these proofs. Here K is a quantity of fixed size thrown in to represent the overhead in the program for things like reading in the number N, communicating with the printer and so on. With this program we can search through all proofs of length 1, length 2 and so on until we come to the one that proves that the complexity of a specific number is greater than N. When such a proof is found, the program of length $\log N + K$ will have generated a string of complexity greater than N. But there will always be some number N such that N is much larger than $\log N + K$, since K is fixed. Thus, on the one hand we have computed a string of complexity N with a program having a length much shorter than N. On the other hand, we have proved that the string has complexity greater than N, which by definition can only be computed with a program of length greater than N—a contradiction. Thus, we conclude that there can be no such proof.

The arguments linking Chaitin's Theorem, Gödel's Theorem and chaos were first presented in

Casti, J. "Chaos, Gödel and Truth," in *Beyond Belief: Randomness, Explanation and Prediction in Science,* J. Casti and A. Karlqvist, eds. Boca Raton, Fla.: CRC Press, 1991, 280–327.

By the isomorphism between formal systems and dynamical systems, it should come as no surprise to learn that almost every interesting property of a dynamical system is undecidable. For instance, it's undecidable whether the dynamics is chaotic, whether a trajectory starting from a specific point eventually passes through a given region of state space, whether the equations are integrable and so forth. Ian Stewart's Rapidly Accelerating Computer and its connection with these undecidability results for dynamical systems is explored further in

Stewart, I. "The Dynamics of Impossible Devices." *Nonlinear Science Today* 1 (1991), 8–9.

The article points out that the reason dynamical systems have such strong undecidability properties is that they are so versatile they can model the computational process itself.

Real Brains, Artificial Minds

Turing's proposal of the Imitation Game as an objective test of machine intelligence was first put forward in the pioneering article

Turing, A. "Computing Machinery and Intelligence." *Mind* 59 (1950), 433–460 (reprinted in *Minds and Machines,* A. Anderson, ed. Englewood Cliffs, N.J.: Prentice-Hall, 1964, 4–30).

A good account of the current state of play in the strong AI business is given in the works

The Artificial Intelligence Debate. S. Graubard, ed. Cambridge, Mass.: MIT Press, 1988.

Denning, P. "The Science of Computing: Is Thinking Computable?" *American Scientist* 78 (March–April 1990), 100–102.

Ned Block's argument against the Turing test was first proposed in the article

Block, N. "Psychologism and Behaviorism." *Philosophical Review* 90 (1981), 43–59.

The strengths and (especially) the weaknesses of the Top Down program for strong AI are tallied in

Dreyfus, H. and S. Dreyfus. "Making a Mind Versus Modeling the Brain: Artificial Intelligence Back at a Branchpoint," in *The Artificial Intelligence Debate,* S. Graubard, ed. Cambridge, Mass.: MIT Press, 1988, 15–43.

Bottom Up work on thinking machines via the route of neural networks is considered in

Crick, F. "The Recent Excitement About Neural Networks." *Nature* 337 (January 12, 1989), 129–132.

Minsky, M. *The Society of Mind.* New York: Simon & Schuster, 1986.

Rumelhart, D. and J. McClelland. *Parallel Distributed Processing*. Vols. 1 & 2. Cambridge, Mass.: MIT Press, 1986.

Johnson, G. *Palaces of Memory*. New York: Knopf, 1991.

The debate between the computer scientists and the philosophers on the (de)merits of strong AI is encapsulated in the following article for the educated layman:

"Artificial Intelligence: A Debate." *Scientific American*, January 1990, 19–31.

The phenomenological argument of the Dreyfus brothers against Top Down AI is summarized in the volumes

Dreyfus, H. and S. Dreyfus. *Mind Over Matter*. New York: The Free Press, 1986.

Dreyfus, H. *What Computers* Still *Can't Do*. Cambridge, Mass.: MIT Press, 1992.

John Searle has presented the ins and outs of his Chinese Room argument against strong AI in many forums. Here is a sampling of the best, starting with the original article that sparked off the whole antibehaviorist debate:

Searle, J. "Minds, Brains and Programs." *Behavioral and Brain Sciences* 3 (1980), 417–457.

Searle, J. "Cognitive Science and the Computer Metaphor," in *Understanding the Artificial,* M. Negrotti, ed. London: Springer, 1991, 127–138.

Searle, J. *Minds, Brains and Science*. Cambridge, Mass.: Harvard University Press, 1984.

Searle, J. *The Rediscovery of the Mind*. Cambridge, Mass.: MIT Press, 1992.

Appeals to Gödel's Theorem as the basis for an anti-AI position have also filled the literature. The most prominent such arguments are the original position taken by Lucas, together with its recent reincarnation by Roger Penrose. The two sources are

Lucas, J. "Minds, Machines and Gödel." *Philosophy* 36 (1961), 120–124 (reprinted in *Minds and Machines*. A. Anderson, ed. Englewood Cliffs, N.J.: Prentice-Hall, 1964, 43–59).

Penrose, R. *The Emperor's New Mind*. Oxford: Oxford University Press, 1989.

Minds, Machines and Evolution

Tom Ray's remarkable work showing the independence of the most important aspects of Darwinian evolution from any particular material substrate has been reported in many places. Two of the most informative are

Ray, T. "An Approach to the Synthesis of Life," in *Artificial Life—II,* C. Langton et al, eds. Redwood City, Calif.: Addison-Wesley, 1992, 371–408.

Maynard Smith, J. "Byte-Sized Evolution." *Nature* 355 (February 27, 1992), 772–773.

It's of more than passing interest that the same line of attack has been followed in the evolutionary "growth" of computer programs aimed at solving definite classes of problems. For an account of these matters, see

Emergent Computation. S. Forrest, ed. Cambridge, Mass.: MIT Press, 1991.

Koza, J. *Genetic Programming*. Cambridge, Mass.: MIT Press, 1992.

Gödel's statement about thinking machines is reported in the article

Rucker, R. "Towards Robot Consciousness." *Speculations in Science and Technology* 3 (1980), 205–217.

There's almost no movement in modern science that's unfolding at a more rapid pace than the development of artificial life. The "bibles" of the field are the published proceedings from the first two international meetings on the topic, both held in Santa Fe, New Mexico, in 1987 and 1989. The references are

Artificial Life. C. Langton, ed. Redwood City, Calif.: Addison-Wesley, 1989.

Artificial Life—II. C. Langton et al., eds. Redwood City, Calif.: Addison-Wesley, 1992.

A popular account of the historical development of the field is entertainingly presented in

Levy, S. *Artificial Life*. New York: Pantheon, 1992.

Steen Rasmussen's postulates underpinning the belief in such a thing as an artificial organism were first reported in

Rasmussen, S. "Aspects of Information, Life, Reality and Physics," in *Artificial Life—II*, C. Langton et al., eds. Redwood City, Calif.: Addison-Wesley, 1992, 767–773.

For a discussion of the parallels between the research programs of the A-lifers and the strong AIers, see

Sober, E. "Learning from Functionalism: Prospects for Strong Artificial Life," in *Artificial Life—II*, C. Langton et al., eds. Redwood City, Calif.: Addison-Wesley, 1992, 749–765.

CHAPTER FIVE

Getting It Together

The quote by Ross Ashby, in which he points out the futility of trying to study complex systems by varying one factor at a time, is taken from the following book, which is about as good an introduction to the world of systems as any I know of:

Ashby, W. Ross. *An Introduction to Cybernetics*. London: Chapman & Hall, 1956.

The best possible accounts of the use of binary relations among sets to describe the interconnections in systems are:

Atkin, R. *Mathematical Structure in Human Affairs*. London: Heinemann, 1974.

Atkin, R. *Multidimensional Man*. London: Penguin, 1981.

Both of these volumes pack an amazing variety of ideas and examples into a small amount of space, giving the reader a heady overview of how to use mathematical concepts in areas usually thought to be far outside the realm of the rational and scientific.

In this same regard, see the works

Casti, J. *Connectivity, Complexity and Catastrophe in Large Systems*. Chichester, England: John Wiley & Sons, 1979.

Gould, P. "Q-Analysis, or a Language of Structure: An Introduction for Social Scientists, Geographers and Planners." *International Journal of Man-Machine Studies* 13 (1980), 169–199.

An up-to-date summary of the state of play vis-à-vis the uses of q-analysis to study complex systems is found in

Johnson, J. "The Mathematics of Complex Systems," in *The Mathematical Revolution Inspired by Computing,* J. Johnson and M. Loomes, eds. Oxford: Oxford University Press, 1991, 165–186.

The example using sets and relations to analyze the Middle East crisis is taken from unpublished work by the author and Mel Shakun of New York University. Some of it is described in more detail in Chapter Eight of

Casti, J. *Reality Rules: Picturing the World in Mathematics. II—The Frontier*. New York: Wiley, 1992.

Making Connections

For the chess aficionado, Chapter Three of the first Atkin book cited above is must reading. For the more technical aspects of how to employ q-analysis ideas to develop playing strategies, see the article

Atkin, R. and I. Witten. "A Multi-dimensional Approach to Positional Chess." *International Journal of Man-Machine Studies* 7 (1975), 727–750.

The number of disjoint chains of q-connection in a complex give us a picture of its overall geometry. However, these chains don't tell us much about how any individual simplex is integrated into the overall structure. For this, we need what R. H. Atkin has called the "eccentricity" of a simplex. If we let $\hat{q}$ be the geometric dimension of a simplex σ, and denote by $\bar{q}$ the lowest-dimensional face that σ shares with any other simplex in the complex, then Atkin defines the eccentricity of σ to be

$$\text{ecc}\ \sigma = \frac{\hat{q} - \bar{q}}{\bar{q} + 1}$$

The rationale underlying this measure is that σ is more "idiosyncratic" within the complex if it has many vertices that it does not share with any other simplex. But this difference is presumably more significant at lower-dimensional levels than at higher, so we normalize by dividing through by the quantity $\bar{q}$. In particular, if σ shares *no* vertices with any other simplex, then $\bar{q} = -1$, leading to an infinite eccentricity.

The Time of Your Life

One of the eeriest precognitive visions of the *Titanic*'s fate was the 1898 novel *Futility,* by the American writer Morgan Robertson. The story line centered around the sinking of a supposedly unsinkable giant of a ship named *Titan.* In the novel, the ship meets her downfall in the North Atlantic, during the month of April, via collision with a large iceberg. And as if this were not enough, the fine-grained details match as well. For example, the *Titan* was traveling at 25 knots at the moment of collision, the *Titanic* at 23 knots; the *Titan* carried 3,000 passengers and crew, a large proportion of whom were lost because the ship had too few lifeboats on board. And so on and so forth. In short, way too much detail to be chalked up to mere coincidence. For further details and examples of this sort of glimpse into the future, including an account of Mrs. Marshall, see

Zohar, D. *Through the Time Barrier.* London: Heineman, 1982.

Trying to get a handle on time has been a human preoccupation at least as long as people have been measuring things. And the puzzle looks no closer to being solved today than when St. Augustine made his famous remark to the effect that he understood perfectly well what time was so long as no one asked him. Relatively recent works addressing what we think we know of time today, scientifically speaking, include

The Nature of Time. R. Flood and M. Lockwood, eds. Oxford: Basil Blackwell, 1986.

The Enigma of Time. P. Landsberg, ed. Bristol, England: Adam Hilger, 1982.

Shallis, M. *On Time.* London: Burnett Books, 1982.

For more information about the multidimensional concept of time discussed in the text, see the Atkin volume *Multidimensional Man* cited above, as well as the article

Atkin, R. "Time as a Pattern on a Multi-dimensional Structure." *Journal of Social & Biological Structures* 1 (1978), 281–295.

Some Surprising Connections

For a layman's account of the quantum measurement problem, see Chapter Seven of

Casti, J. *Paradigms Lost.* New York: Morrow, 1989 (paperback edition: Avon Books, New York, 1990).

The connection given in the text between classical probability theory and simplicial complexes is discussed at greater lengths in the Atkin volume *Multidimensional Man,* cited earlier. See also Appendix C of Atkin's *Mathematical Structures in Human Affairs* for a worked-out example of a situation in which the complex associated with a die-tossing experiment has "holes."

This example shows the utility of thinking of probability theory in multidimensional terms.

As promised in the text, here is Atkin's measure of the surprise of a q-event σ_q relative to a p-dimensional base event σ_p^0:

$$\text{surp}(\sigma_q \bmod \sigma_p^0) = \frac{n_q(\sigma_q, \sigma_p^0)}{p+1}$$

where $n_q(\sigma_q, \sigma_p^0)$ is the number of disjoint q-chains linking the event in question σ_q with the base event σ_p^0. The satisfying fact is that this measure of surprise agrees with the ordinary probability of an event—provided that all events are 0-events! For a detailed discussion of these connections and much, much more, see the pioneering article

Atkin, R. "A Theory of Surprises." *Environment and Planning B* 8 (1981), 359–365.

CHAPTER SIX

Checkerboard Computers

By all accounts, cellular automata seem to have first been developed by mathematician Stanislaw Ulam in the late 1940s or thereabouts. When von Neumann ran into difficulties with his initial attempts at solving the self-reproduction problem, primarily on account of the mechanical complications inherent in the scheme he was using, Ulam suggested a purely information-theoretic approach to the question via CAs. The rest, as they say, is history. A layman's introduction to this circle of ideas can be found in the volume

Levy, S. *Artificial Life.* New York: Pantheon, 1992.

Good introductions to cellular automata from a computational point of view are

Toffoli, T. and N. Margolus. *Cellular Automata Machines.* Cambridge, Mass.: MIT Press, 1987.

Cellular Automata: Theory and Experiment. H. Gutowitz, ed. Cambridge, Mass.: MIT Press, 1991.

For a semitechnical introduction to the entire field, along with applications of CA in biology, languages and economics, see Chapter Three of

Casti, J. *Reality Rules: Picturing the World in Mathematics. I—The Fundamentals.* New York: Wiley, 1992.

Tom Schelling's experiments on racial integration in urban housing are reported in

Schelling, T. "Dynamic Models of Segregation." *Journal of Mathematical Sociology* 1 (1971), 143–186.

During the 1960s, CA fell out of favor with applied mathematicians, but in the 1970s, work led by Stephen Wolfram revitalized the field. Many of

Wolfram's pioneering efforts, along with those of other researchers, are reported in the volume

Theory and Applications of Cellular Automata. S. Wolfram, ed. Singapore: World Scientific Press, 1986.

Aristid Lindenmayer died in 1989. But his work on L-systems has been carried on by his many students and co-workers. For an excellent summary of these efforts, complete with a stunning collection of color plates displaying a collection of plants seen only in the world inside the computer, the reader should see the volume

Prusinkiewicz, P. and A. Lindenmayer. *The Algorithmic Beauty of Plants*. New York: Springer, 1990.

That's Life?

Von Neumann's proof of the possibility of self-reproducing automata is given in

Von Neumann, J. *Theory of Self-Reproducing Automata*. Urbana, Ill.: University of Illinois Press, 1966.

Unfortunately, this original work is rather difficult to follow. Simpler accounts of von Neumann's ideas are presented from several points of view in the collection

Essays on Cellular Automata. A. Burks, ed. Urbana, Ill.: University of Illinois Press, 1970.

The idea of a living organism as a machine has proven irresistibly attractive to scientists and philosophers since the time of Aristotle. Some more recent perspectives on this eternal question are found in the papers

Laing, R. "Machines as Organisms: An Exploration of the Relevance of Recent Results." *Biosystems* 11 (1979), 201–215.

Laing, R. "Anomalies of Self-Description." *Synthese* 38 (1978), 373–387.

As noted in the text, Conway's *Life* game was brought to the attention of the general public in a series of articles by Martin Gardner in *Scientific American*. The complete set of *Life* articles, as well as a number of related topics, can be found in

Gardner, M. *Wheels, Life and Other Mathematical Amusements*. San Francisco: W. H. Freeman, 1983.

An informative, yet popular account of the *Life* game, together with computer programs for playing it, is given in

Poundstone, W. *The Recursive Universe*. New York: Morrow, 1985.

This volume is also noteworthy for its extended discussion of the question of self-reproducing *Life* patterns.

Von Neumann's self-rep work, the *Life* game and L-systems formed the precursors to what today is the thriving field of "artificial life." Devotees are united in the belief that what distinguishes living things from the nonliving is their functional organization, not their material form. The artificial lifer's "bible" is the volume containing the proceedings of the historic 1987 Los Alamos conference, which brought the various strands of the A-life community together for the first time. This work contains reports on theoretical aspects of self-reproduction and what it means to be "alive," as well as many accounts of artificial lifeforms that are currently cavorting about in the memory banks of computers across America and around the world. Operating on the premise that one good workshop deserves another, the Santa Fe Institute organized a second A-life workshop in February 1990 that continued the tradition established in the path-breaking 1987 event. So for anyone wanting to know about AL, the Levy book cited at the beginning of this chapter, as well as the two proceedings volumes from these A-life workshops, are the places to look. The precise citations are for the workshop volumes are

Artificial Life. C. Langton, ed. Redwood City, Calif.: Addison-Wesley, 1989.

Artificial Life—II. C. Langton et al, eds. Redwood City, Calif.: Addison-Wesley, 1992.

Another introductory volume containing much information on the goings-on in the A-life world, together with a computer diskette for performing a few experiments of your own, is

Rietman, E. *Creating Artificial Life*. Blue Ridge Summit, Penn.: Windcrest/McGraw-Hill, 1993.

The Most Complicated Thing in the World

The theory and application of fractals has become almost a cottage industry by now, with the number of volumes on the topic, not to mention computer programs, threatening to swamp scientific booksellers' shelves. Here are just a few of the better places to look for introductory accounts of the coastline of Britain, Sierpinski gaskets and carpets, Julia sets and a whole lot more, starting with Mandelbrot's classic work that sparked off the subject:

Mandelbrot, B. *The Fractal Geometry of Nature*. San Francisco: W. H. Freeman, 1982.

Peitgen, H.-O., D. H. Jürgens and D. Saupe. *Fractals for the Classroom*. Parts 1 & 2. New York: Springer, 1992.

Schroeder, M. *Fractals, Chaos, Power Laws*. New York: W. H. Freeman, 1991.

Feder, J. *Fractals*. New York: Plenum, 1988.

Mandelbrot, B. "On Fractal Geometry, and a Few of the Mathematical Questions It Has Raised." *Proceedings of the International Congress of Mathematicians,* Warsaw, 1983, 1661–1675.

Just as with the catastrophe theory brouhaha a few years ago, fractal geometry is not without its naysayers, claiming that fractals have neither answered any old questions nor asked any new ones. Geometer Stephen Krantz of Washington University in St. Louis is one of the more outspoken antifractalists. He has not only attacked the subject itself but for good measure has thrown in a few ad hominem attacks on Mandelbrot himself. A summary of these skirmishes in the trenches of science is found in

Bown, W. "Twisting the Fractal Knife." *New Scientist,* September 29, 1990, 63.

A layman's account of why the Mandelbrot set is the most complicated thing in the world is given in the article

Bown, W. "Mandelbrot Set Is as Complex as It Could Be." *New Scientist,* September 28, 1991, 22.

From Bach to Rock and Bach Again

Frequencies in time and space that appear to be random on any scale underlie many natural processes, leading to the so-called 1/f noise discussed in the text. For a good introductory account of these matters, see the article

West, B. and M. Schlesinger. "The Noise in Natural Phenomena." *American Scientist* 78 (January–February 1990), 40–45.

A good account of Zipf's Law is found in the volume

Nicolis, J. *Chaos and Information Processing.* Singapore: World Scientific, 1991.

The discussion of fractal music follows that given in the Schroeder book above, as well as the journalistic account

Browne, M. "J. S. Bach + Fractals = New Music." *The New York Times,* April 16, 1991.

See also the article

"White, Brown and Fractal Music," in Gardner, M. *Fractal Music, Hypercards, and More.* New York: W. H. Freeman, 1992.

Climbing the Devil's Staircase

For more details on the relative merits of timid versus bold play in the casinos, see the Schroeder and Feder volumes cited earlier.

A summary of Hurst's studies is given in the Feder volume. For the complete story, see Hurst's original book,

Hurst, H., R. Black and Y. Simaika. *Long-Term Storage: An Experimental Study.* London: Constable, 1951.

The world's financial markets have always been a source of attraction for scientists in the grip of a new theory, with the current explosion of interest in chaos and fractals proving no exception. Most of the work reported in the text is covered in the thought-provoking work

Peters, E. *Chaos and Order in the Capital Markets.* New York: Wiley, 1991.

For an introductory account of the ins and outs of the stock market, including the conventional academic wisdom of efficient markets and what's wrong with them, see Chapter Four of

Casti, J. *Searching for Certainty.* New York: Morrow, 1991 (paperback edition: Quill, New York, 1992).

William Brock, Blake LeBaron and their colleagues at the University of Wisconsin have led the charge insofar as statistical investigations of chaotic behavior in speculative markets goes. A good summary of this work is reported in

Brock, W., D. Hsieh, and B. LeBaron. *Nonlinear Dynamics, Chaos, and Instability: Statistical Theory and Economic Evidence.* Cambridge, Mass.: MIT Press, 1991.

Brock, W. "Chaos and Complexity in Economic and Financial Science," in *Acting Under Uncertainty: Multidisciplinary Conceptions,* G. von Furstenberg, ed. Dordrecht, Netherlands: Kluwer, 1990, Chapter 17.

Brock, W. "Causality, Chaos, Explanation and Prediction in Economics and Finance," in *Beyond Belief: Randomness, Explanation and Prediction in Science,* J. Casti and A. Karlqvist, eds. Boca Raton, Fla.: CRC Press, 1990, 230–279.

CHAPTER SEVEN

The Anatomy of Surprise

The Arrow Impossibility Theorem is probably the central result in what's normally called social choice theory. It addresses the basic question How can many individuals' preferences be combined to yield a collective choice? The theorem says, in essence, that there is no way to combine the individual preferences so that everyone will be satisfied. One of the best sources for a thorough discussion of Arrow's Theorem is the work

MacKay, A. *Arrow's Theorem: The Paradox of Social Choice.* New Haven: Yale University Press, 1980.

The discussion of Prisoner's Dilemma situations and sports drafts follows that given in the article

Brams, S. and P. Straffin. "Prisoners' Dilemma and Professional Sports Drafts." *American Mathematical Monthly* 86 (1979), 80–88.

Here is an example to illustrate the point that when three or more teams are involved, it's possible for a Prisoners Dilemma situation to arise. Suppose there are three teams, A, B and C, with six players in the draft, 1–6. Suppose the teams have the following preference orderings: Team A: $1 > 2 > 3 > 4 > 5 > 6$, Team B: $5 > 6 > 2 > 1 > 4 > 3$, Team C: $3 > 6 > 5 > 4 > 1 > 2$. Assume the teams use sophisticated play and that there are two rounds, resulting in the selections given in the following table:

	Team A	*Team B*	*Team C*
Round 1	3	5	6
Round 2	1	2	4

However, if the teams had made sincere choices, they would have ended up with the following allocation of players:

	Team A	*Team B*	*Team C*
Round 1	1	5	3
Round 2	2	6	4

Comparison of these two allocations shows that all three teams are worse off by following sophisticated play than they would have been had they chosen sincerely.

More information on the use of increasing returns to study industry agglomeration is given in the article

Arthur, W. B. "'Silicon Valley' Locational Clusters: When Do Increasing Returns Imply Monopoly?" *Mathematical Social Sciences* 19 (1990) 235–251.

Results showing the uncomputability of solutions of the wave equation and many other important quantities in mathematics and physics are treated in the volume

Pour-El, M. and J. Richards. *Computability in Analysis and Physics*. Berlin: Springer Verlag, 1989.

A layman's account of the five-particle system by which Z. Xia showed the difficulty of the N-Body Problem is given in

Stewart, I. "Cosmic Tennis Blasts Particles to Infinity." *New Scientist,* October 3, 1992, 14.

Stuart Kauffman's work on genetic networks is described in detail in his many books and articles. Good summaries are given in

Kauffman, S. *Origins of Order*. Oxford: Oxford University Press, 1992.

Kauffman, S. "Origins of Order in Evolution: Self-Organisation and Selection," in *Theoretical Biology,* B. Goodwin and P. Saunders, eds. Edinburgh: Edinburgh University Press, 1989, 67–88.

"Complexification"

For somewhat more detailed accounts of what it could be like to be a complex system, the reader should consult the potpourri of ideas in the following works:

Complexity, Language and Life. J. Casti and A. Karlqvist, eds. Berlin: Springer Verlag, 1986.

Lloyd, S. "The Calculus of Intricacy." *The Sciences,* September–October 1990, 38–44.

Lectures in the Sciences of Complexity. D. Stein, ed. Redwood City, Calif.: Addison-Wesley, 1989.

Casti, J. "The Simply Complex: Trendy Buzzword or Emerging New Science?" *Bulletin of the Santa Fe Institute* 7, no. 1 (Spring–Summer 1992) 10–13.

The Science of Surprise

It's difficult to think of a field more fashionable nowadays than complex systems. But as noted in the text, the term *complexity* seems to mean many different things to many different people. The idea of an actual *science* of complexity is one currently being pursued at many research centers throughout the world, especially at the Santa Fe Institute in New Mexico. For lively accounts of the origin and development of this unique institution, see the popular volumes

Waldrop, M. *Complexity*. New York: Simon & Schuster, 1992.

Lewin, R. *Complexity*. New York: Macmillan, 1992.

Shorter introductory accounts are also given in supplements to *New Scientist* magazine in its issues of February 6 and 13, 1993.

The modeling relation of Figure 7.1 is the diagram expressing the essence of what it means to do theoretical science. An extended discussion of this diagram and its many implications for both physics and biology is found in:

Rosen, R. *Life Itself*. New York, Columbia University Press, 1991.

Further discussion of the idea that it is the number of inequivalent descriptions of a system that determines its complexity is given in Chapter One of

Casti, J. *Reality Rules: Picturing the World in Mathematics. I—The Fundamentals*. New York: Wiley, 1992.

INDEX

Italic numbers refer to figures; bold page numbers refer to tables.

CREDITS

Grateful acknowledgment is made to the following individuals and publishers for permission to reproduce material used in creating the figures in this book. Every effort has been made to locate the copyright holders of material used here. Omissions brought to our attention will be corrected in future editions.

Academic Press for Figures 2.1, 2.2, 2.16 and 2.17, which are reproduced from *Transformations,* C. Renfrew and K. Cooke, eds., 1979.

John Wiley & Sons for Figures 3.11, 3.12, 6.30, 6.31 and 6.32, which are reproduced from Peters, E. *Chaos and Order in the Capital Markets,* 1991.

Cambridge University Press for Figures 2.6 and 2.10–2.15, which are reproduced from Saunders, P. *An Introduction to Catastrophe Theory,* 1980, and for Figures 3.1 and 3.2, which are reproduced from Jackson, A. *Perspectives of Nonlinear Dynamics,* Vol. 2, 1991.

New Scientist magazine for Figures 1.1, 3.3–3.7 and 3.10.

Mitchell Beazley International, Ltd. for Figure 1.12, which is reproduced from *The American Express Guide to New York,* 1986.

E. P. Dutton, Inc. for Figures 2.3, 2.7 and 2.8, which are reproduced from Woodcock, T. and M. Davis. *Catastrophe Theory,* 1978.

Penguin Books for Figures 5.2, 5.5, 5.7 and 5.12, which are reproduced from Atkin, R. H. *Multidimensional Man,* 1981.

Houghton Mifflin Company for Figures 4.4 and 4.6, which are reproduced from Rucker, R. *Mind Tools,* 1987.

University of Chicago Press for Figures 3.8 and 3.9, which are reproduced from Ekeland, I. *Mathematics and the Unexpected,* 1988.

OMNI magazine for Figure 1.2.

W. H. Freeman & Company for Figure 1.4, which is reproduced from Peterson, I. *Islands of Truth,* 1990, for Figures 1.6 and 6.19–6.21, which are reproduced from Peterson, I. *The Mathematical Tourist,* 1988, and for Figures 6.12, 6.23–6.25 and 6.29, which are reproduced from Schroeder, M. *Chaos, Fractals, Power Laws,* 1991.

Birkhäuser Boston for Figure 1.5, which is reproduced from Arnold, V. *Huygens and Barrow, Newton and Hooke,* 1990.

Heinemann Publishing Company for Figures 1.7, 5.9 and 5.10, which are reproduced from Atkin, R. H. *Mathematical Structure in Human Affairs,* 1974.

Doubleday Publishing Company for Figures 1.8, 1.9 and 4.2, which are reproduced from Faletta, N. *The Paradoxicon,* 1983.

Nature magazine for Figures 1.10 and 1.11.

Viking Press for Figure 1.15, which is reproduced from Gleick, J. *Chaos,* 1987.

Springer Verlag, Inc. for Figures 6.10, 6.11 and 6.14–6.18, which are reproduced from Peitgen, H.-O., et al. *Fractals for the Classroom, Part One,* 1992 and for Figure 6.22, which is reproduced from Peitgen, H.-O. and P. Richter. *The Beauty of Fractals,* 1986.

Plenum Publishing Corporation for Figures 6.27 and 6.28, which are reproduced from Feder, J. *Fractals,* 1988.

The New York Times for Figure 6.26.

James Crutchfield for Figure 1.14.

Cesaré Marchetti for Figures 1.18–1.20.

E. C. Zeeman for Figures 2.4, 2.5 and 2.9.

Ian Stewart for Figure 3.4.

Erik Mosekilde for Figures 3.13–3.15

Jeff Johnson for Figure 5.6.

Thomas C. Schelling for Figures 6.1 and 6.2.